AF822364

2024

UNIFORM PLUMBING CODE

STUDY GUIDE™

IAPMO®

First Printing, March 2024

Published by the International Association of Plumbing and Mechanical Officials
4755 E. Philadelphia Street • Ontario, CA 91761-2816 – USA
Main Phone: (909) 472-4100 • Main Fax: (909) 472-4150

INTRODUCTION

This Study Guide was developed as a tool for self-study or classroom training to assist the user to better understand the requirements of the Uniform Plumbing Code (UPC). Although it is quite useful for examination preparation, such as, inspector certification, contractor, and journeyman, it is not intended to replace the intense study required to be successful for these exams. This book is not intended to replace the UPC document.

To get the most benefit from this book, read the beginning paragraph of each study guide chapter. If you are preparing for an open book exam, read the question and answer choices, then attempt to locate the subject matter in the UPC as quickly as possible. Proceed in this manner until you have completed the exam. Verify your answers in the key located in the back of this book. If you are preparing for a closed book exam, read each question and answer carefully, then choose the correct answer. Check your work using the answer key located in the back.

When you have completed the practice exams for each chapter, you will be ready to attempt the general exams located after the study guide chapters. Note: If you are taking a timed exam, be sure to give yourself the same amount of time to complete the general exams.

In addition to the three general exams this book contains a math exam and trade exams to further test your knowledge. Technical drawings of buildings and mechanical systems are provided where necessary. The trade exam pertaining to gas systems includes piping diagrams for additional sizing.

IMPORTANT NOTE: Although some questions may appear similar to those you see on your exam, this book is not based on any specific type or classification of exam. No exams were used to write this book; therefore, any questions or answers that appear to be identical to any certification or licensing exam is purely coincidental and accidental.

The Publication Development Committee is committed to continuously providing the plumbing and mechanical industry with the most extensive certification exams available. These exams are designed to authenticate the examinees' integrity and knowledge and understanding of codes, with the end goal of helping to ensure the public's health and safety. Our dedication to the development and improvement of the UPC and UMC Study Guides will provide the industry with unsurpassed study resources that support IAPMO's commitment to its educational goals and objectives.

The 2023 Publication Development Committees shared in the development of and final approval of this publication. Since that time, changes in the committee membership may have occurred.

2023 IAPMO
PUBLICATION DEVELOPMENT COMMITTEE

Martin Cooper, *Chairman*
City of Foster City

Cristina Barillas-McEntee, Plumbers Local 130
Mark Berdichevshy, San Francisco, CA
Kenneth Borski, City of Houston/Planning & Dev.
Michael Durfee, County of Salt Lake
Jason Fernandes, UA Local 367
Shannon Ramey, Jensen Hughes
Matthew Schmidt, City of Lincoln
Jeremy Stettler, Davis School District, Utah
Daniel Cole, IAPMO Staff Liaison
Stu Asay, IAPMO Staff

TABLE OF CONTENTS

CHAPTER 1
ADMINISTRATION

The questions found in this chapter are designed to test your knowledge in the enforcement requirements of the Uniform Plumbing Code. These requirements include the duties and powers of the Authority Having Jurisdiction, permit issuance, inspections, and exempt work, along with requirements for existing plumbing systems.

1. This document shall be known as the _____.
   (A) Uniform Plumbing Code
   (B) Universal Plumbing Code
   (C) National Plumbing Code
   (D) World Plumbing Code

2. The _____ shall be permitted to require the submission of plans, specifications, drawings, and such other information, prior to the commencement of work regulated by this code.
   (A) plumbing inspector
   (B) building authority
   (C) Authority Having Jurisdiction
   (D) building official

3. The provisions of _____ shall apply to the erection, installation, alteration, repair, relocation, replacement, addition to, use, or maintenance of plumbing systems within this jurisdiction.
   (A) this code
   (B) the Building Code
   (C) the International Code
   (D) NFPA 99C

4. All openings into a drainage or vent system, except those openings to which plumbing fixtures are properly connected or that constitute vent terminals, shall be permanently plugged or capped _____ in accordance with this code.
   (A) with old rags
   (B) with duct tape
   (C) using appropriate materials
   (D) with plastic tape

5. The _____ shall be responsible for maintenance of plumbing systems.
   (A) city
   (B) plumbing inspector
   (C) owner or owner's agent
   (D) building department

6. Plumbing systems, lawfully in existence at the time of the adoption of this code shall be permitted to have their use, maintenance, or repair continued as long as it is in accordance with the original design and no hazard to _____ has been created by such plumbing system.
   (A) the owner
   (B) a tenant
   (C) the city
   (D) life, health, or property

7. When the requirements within the jurisdiction of this plumbing code conflict with the requirements of the mechanical code, _____ shall prevail.
   (A) the Mechanical Code
   (B) this code
   (C) the Building Code
   (D) either code

8. Plumbing systems that are a part of a building or structure undergoing a change in use or occupancy, as defined by the building code, shall be in accordance with the requirements of _____ that are applicable to the new use or occupancy.
   (A) the code in effect at time of construction
   (B) this code
   (C) the Building Code
   (D) the plumbing inspector

9. Parts of the plumbing system of a building or part thereof that is moved from one foundation to another, or from one location to another, shall be in accordance with the provisions of ______ for new installations.
   (A) the building code
   (B) the code in effect when originally constructed
   (C) this code
   (D) any code

10. The ______ shall be the authority duly appointed to enforce this code.
    (A) Authority Having Jurisdiction
    (B) plumbing inspector
    (C) building official
    (D) building inspector

11. Where work is being done contrary to the provisions of this code, the Authority Having Jurisdiction shall be permitted to order the ______ by notice in writing served on persons engaged in the doing or causing such work to be done.
    (A) completion of work
    (B) immediate inspection
    (C) work stopped
    (D) summons to court

12. It shall be unlawful for any person, firm, or corporation to make any installation, alteration, repair, replacement, or remodel a plumbing system regulated by this code, except as permitted in section 104.2, or to cause the same to be done without first obtaining ______ for each separate building or structure.
    (A) permission
    (B) authorization
    (C) a separate plumbing permit
    (D) a work order from owner

13. To obtain a permit, the applicant shall first file ______ in writing on a form furnished by the Authority Having Jurisdiction for that purpose.
    (A) a request
    (B) for permission
    (C) an application
    (D) plans

14. Plumbing systems for which a permit is required by this code shall be inspected by the ______.
    (A) building inspector
    (B) Authority Having Jurisdiction
    (C) mechanical inspector
    (D) health department

15. Plumbing systems shall be ______ in accordance with this code or the Authority Having Jurisdiction.
    (A) inspected
    (B) installed
    (C) tested and approved
    (D) permitted

16. It shall be the duty of the ______ authorized by a permit to notify the Authority Having Jurisdiction that such work is ready for inspection.
    (A) person doing the work
    (B) plumber
    (C) owner
    (D) general contractor

17. The Authority Having Jurisdiction shall be notified that work is ready for inspection not less than ______ hours before the work is to be inspected.
    (A) 72
    (B) 24
    (C) 48
    (D) 12

18. It shall be the duty of the ______ to make sure that the work will stand the test prescribed before giving the notification for inspection.
    (A) owner
    (B) general contractor
    (C) person doing the work
    (D) holder of the permit

19. No person shall make connections from a source of energy or fuel to any plumbing system or equipment regulated by this code and for which a permit is required until approved by the _____.
    (A) gas company
    (B) Authority Having Jurisdiction
    (C) owner
    (D) contractor

20. This code is an ordinance to provide _____ for the protection of the public health, safety, and welfare.
    (A) guidelines
    (B) maximum requirements and standards
    (C) suggested ways of installation
    (D) minimum requirements and standards

21. Where a section, subsection, sentence, clause, or phrase of this code is, for a reason held to be unconstitutional, such decision shall _____.
    (A) render the entire code is invalid
    (B) not affect the validity of the remaining portions of this code
    (C) apply in every jurisdiction in the state
    (D) allow Authority Having Jurisdiction to overrule the court decision

22. Whenever, in this code, reference is made to an appendix, the appendix shall not apply unless formally _____.
    (A) adopted
    (B) stated
    (C) noted
    (D) provided for

23. Temporary connection of the plumbing equipment to the source of energy or fuel for the purpose of testing _____.
    (A) is not allowed unless the inspector is present
    (B) shall be permitted to be authorized by the Authority Having Jurisdiction
    (C) is not allowed unless the equipment has been inspected and approved
    (D) is not allowed under any circumstances

CHAPTER 2
DEFINITIONS

Practical interpretation and application of the Uniform Plumbing Code is possible when all the words and terms used within the code are understood. Since the code is widely and extensively used, the definitions for the words in the code should come from a universal source. For the majority of words and terms in the code, the accepted source is the dictionary. However, there are certain words and terms that have taken on a special meaning because of their specialized use within the plumbing field. Therefore, in order to avoid misunderstanding, the code has given a section specifically for the purpose of defining these special words.

The questions in this examination will test you on your knowledge of these special words and their code definitions. Ordinary words, which are used in accordance with their dictionary meaning, are excluded from this examination. Before taking this examination, you should study all the words and terms that are listed in the code on definitions. It is important that you know these special words and are able to recognize them throughout the code book. A little effort here will pay off in a better insight and understanding of code regulations and requirements.

1. A room equipped with a shower, bathtub or combination bath/shower is a _____.
   (A) toilet room
   (B) bathroom
   (C) shower room
   (D) washroom

2. Fixture connections that require the removal of an access panel for servicing are considered _____.
   (A) accessible
   (B) readily accessible
   (C) not accessible
   (D) exposed

3. That part of the lowest piping of a drainage system that receives the discharge from soil, waste, and other drainage pipes inside the walls of the building and conveys it to a point 2 feet outside the building wall is a _____.
   (A) building drain
   (B) building sewer
   (C) waste pipe
   (D) soil stack

4. A vertical vent that is a continuation of the drain to which it connects is a _____.
   (A) circuit vent
   (B) relief vent
   (C) continuous vent
   (D) branch vent

5. A horizontal drain that is sized to provide free movement of air above the flow line of the drain is called a _____.
   (A) continuous waste system
   (B) durham waste and vent system
   (C) combination waste and vent system
   (D) solvent waste and vent system

6. The length along the centerline of the pipe and fittings is the _____.
   (A) diagonal length
   (B) diameter length
   (C) circumference length
   (D) developed length

7. A device integrated within an air accumulator vessel that is designed to discharge a predetermined quantity of water to fixtures for flushing purposes is a (an) _____.
   (A) expansion tank
   (B) surge tank
   (C) flushometer tank
   (D) day tank

8. The combination of fittings that may be used to make an offset and is considered a vertical pipe are _____.
   (A) 1 - 90 degree and 1 - 45 degree
   (B) 2 - 90 degree and 1 - 45 degree
   (C) 1 - 45 degree and 1 - 60 degree
   (D) 1 - 45 degree and 1 - 45 degree

9. Water that is satisfactory for drinking, culinary and domestic purposes and that meets the requirements of the Authority Having Jurisdiction is referred to by this code as _____.
   (A) potable water
   (B) industrial water
   (C) soft water
   (D) hard water

10. By code definition, a stack is always _____.
   (A) straight
   (B) horizontal
   (C) vertical
   (D) exposed

11. A vent that also serves as a drain is a (an) _____.
   (A) island vent
   (B) vent stack
   (C) loop vent
   (D) wet vent

12. A pipe that makes an angle of not more than 45 degrees with the vertical is considered _____.
   (A) vertical
   (B) horizontal
   (C) crooked pipe
   (D) diagonal pipe

13. The vertical depth of a trap seal is measured from _____.
   (A) the crown weir to the top of the dip of the trap
   (B) the inlet of the trap to the bottom of the trap
   (C) the outlet of the trap to the bottom of the trap
   (D) the inlet of the trap to the outlet of the trap

For questions 14 through 23, select the word or term that fits the sketch.

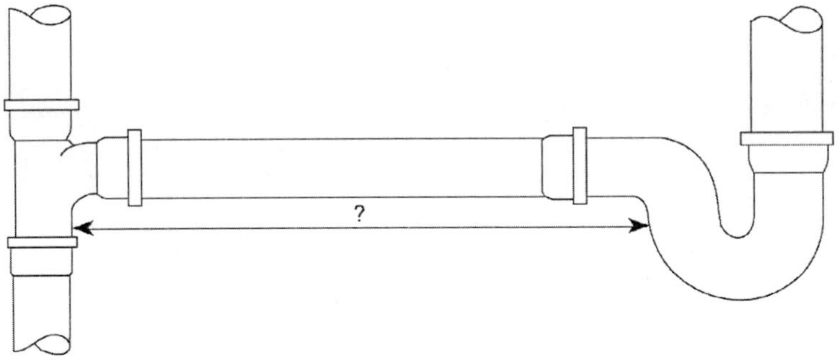

Figure 14

14. (A) waste pipe
   (B) drain pipe
   (C) dirty arm
   (D) trap arm

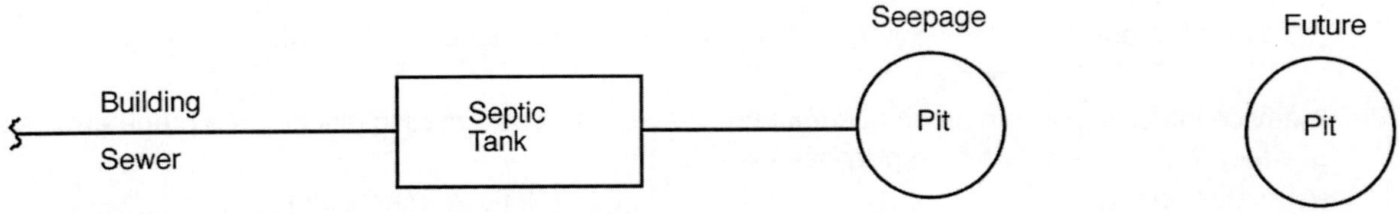

Figure 15

15. (A) industrial waste disposal system
   (B) private sewage retention system
   (C) public sewage retention system
   (D) private sewage disposal system

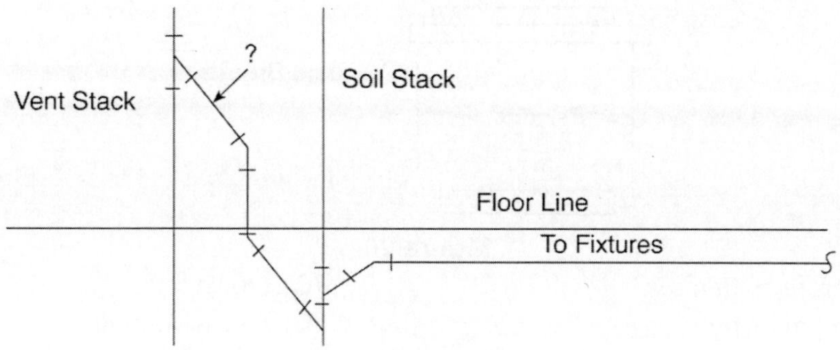

Figure 16

16. (A) relief vent
    (B) yoke vent
    (C) fixture vent
    (D) branch vent

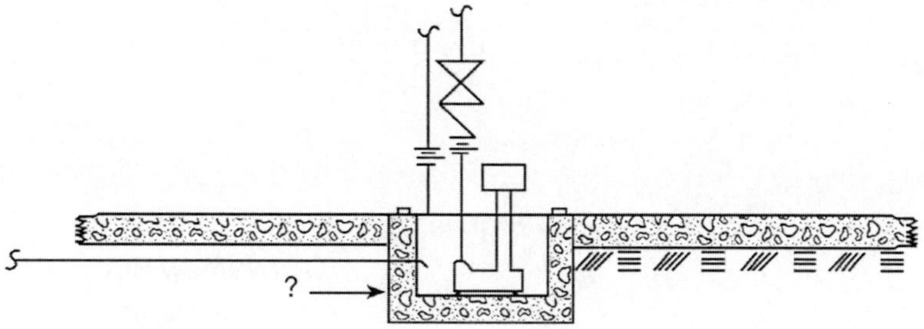

Figure 17

17. (A) holding tank
    (B) catch basin
    (C) sand trap
    (D) sump

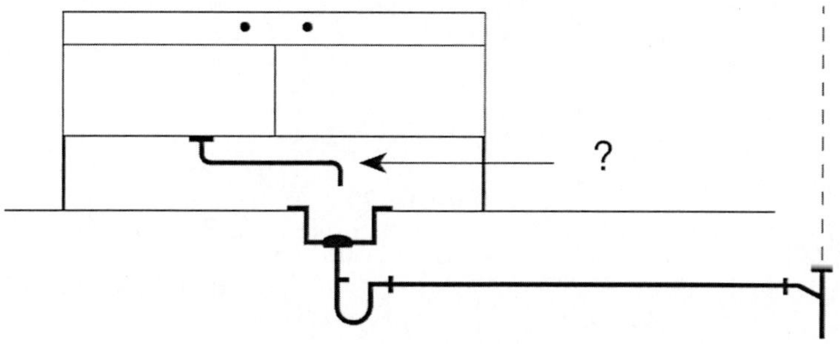

Figure 18

18. (A) fixture branch
    (B) industrial waste
    (C) indirect waste pipe
    (D) island sink drain

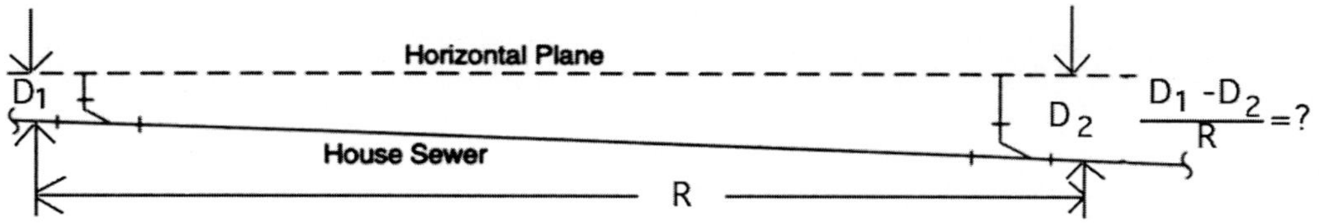

Figure 19

19. (A) distance below grade
    (B) grade of pipe
    (C) velocity of flow
    (D) direction of flow

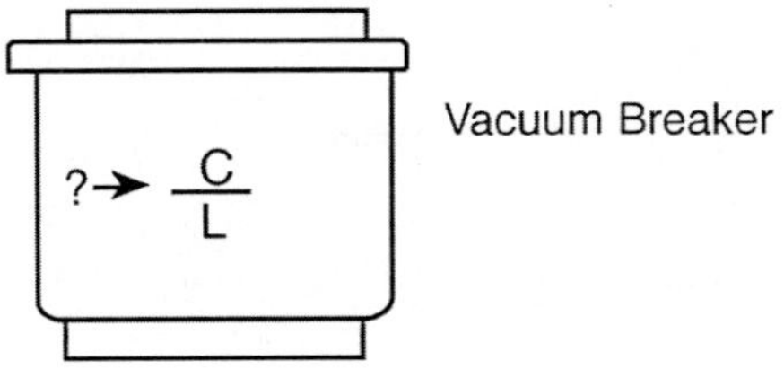

Figure 20

20. (A) manufacturer's trademark
    (B) critical level marking
    (C) crown line
    (D) corrosive limit

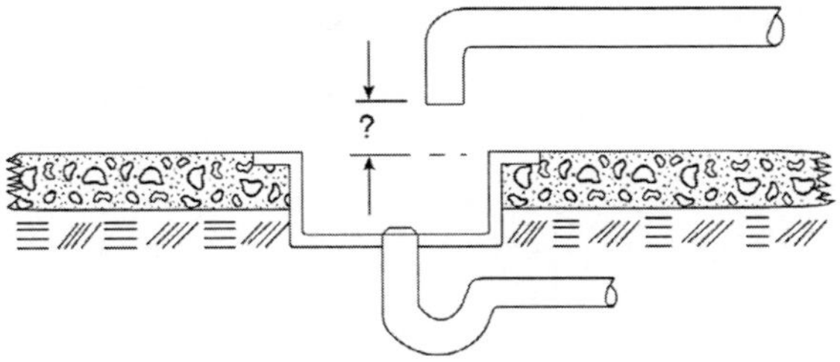

Figure 21

21. (A) airgap
    (B) airbreak
    (C) overflow rim
    (D) static head

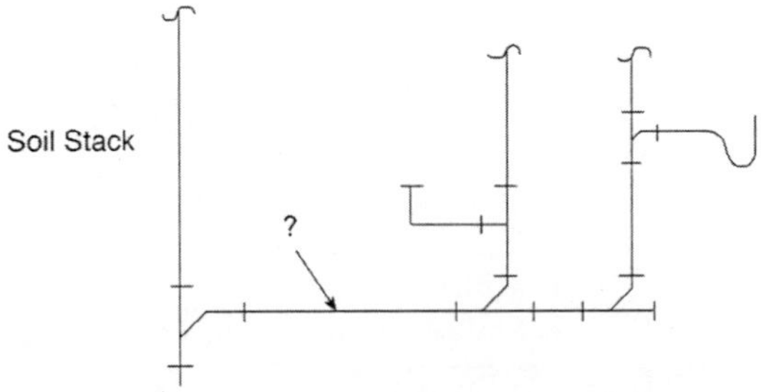

Figure 22

22. (A) fixture supply
    (B) fixture drain
    (C) horizontal branch
    (D) building drain

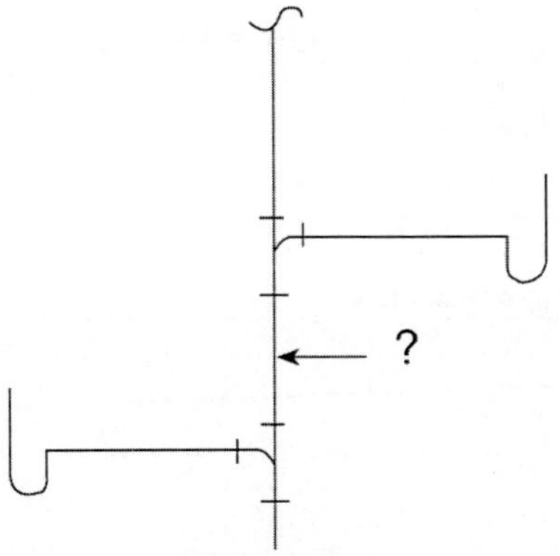

Figure 23

23. (A) individual vent
    (B) wet vent
    (C) wet drain
    (D) yoke vent

24. A mixing valve that senses outlet temperature and compensates for fluctuations in incoming hot or cold water temperature is a _____.
(A) thermostatic (temperature control) valve
(B) combination thermostatic/pressure-balancing valve
(C) pressure-balancing valve
(D) none of the above

25. The restroom facilities in a private day care center are classified as public use.
(A) True
(B) False

26. The pipe receiving the discharge from a clinical sink and conveying it to the building drain is defined as a waste pipe.
(A) True
(B) False

27. A combined building sewer is a _____.
(A) combination of all sewer and drainage piping from inside and outside of a building
(B) building sewer that conveys industrial waste and domestic sewage
(C) building sewer that conveys both sewage and storm water
(D) building drain that conveys waste and storm water

28. PEX is defined as _____.
(A) polyethylene
(B) cross-linked polyethylene
(C) polybutylene X
(D) none of the above

29. The lowest portion of the inside of a horizontal pipe is called the _____.
(A) bottom
(B) flat vent
(C) invert
(D) invert mass

30. The part of the horizontal piping of a drainage system that begins _____ feet downstream from the last mobile home site and conveys it to a public sewer is called a mobile home park sewer.
(A) 2
(B) 3
(C) 3-1/2
(D) 5

31. A device and system of piping that maintains a water seal in a remote trap is called a trap primer
(A) True
(B) False

32. Static pressure is pressure existing without any flow.
(A) True
(B) False

33. A permanently installed mechanical device, other than an ejector, for removing sewage or liquid waste from a sump is called a sewage pump.
(A) True
(B) False

34. A standard that is an extensive compilation of provisions covering broad subject matter is the definition for _____.
(A) standards
(B) listed
(C) legal document
(D) code

35. The principal artery of the venting system to which branches may be connected is called the _____.
(A) branch vent
(B) vertical vent
(C) vent
(D) main vent

36. A joint obtained by the joining of metal parts with metallic mixtures or alloys that melt at a temperature up to and including 840°F is a _____.
    (A) soldered joint
    (B) brazed joint
    (C) metallic joint
    (D) flared joint

37. The organization, office, or individual responsible for enforcing the requirements of a code or standard is the _____.
    (A) administrative authority
    (B) building department
    (C) plumbing inspector or department
    (D) Authority Having Jurisdiction

38. That portion of a drainage system that does not drain by gravity into the building sewer is known as a _____.
    (A) building supply
    (B) building subdrain
    (C) sewer ejector
    (D) storm sewer

39. A pipe inside a building that conveys storm water from the roof to a storm drain, combined building sewer, or other approved point of disposal is called a (an) _____.
    (A) conductor
    (B) building storm drain
    (C) adductor
    (D) rain leader

40. An outlet on a boiler to permit emptying or discharging of sediment is called a (an) _____.
    (A) relief valve
    (B) atmospheric check valve
    (C) aspirator
    (D) boiler blowoff

41. A device designed to provide protection against hydraulic shock in the building water supply system is called a (an) _____.
    (A) backflow preventor
    (B) air chamber
    (C) flushometer valve
    (D) water hammer arrestor

42. A device installed in a drainage system to prevent reverse flow is a _____.
    (A) sewage ejector
    (B) check valve
    (C) backwater valve
    (D) backflow preventor

43. Any pressure less than that exerted by the atmosphere is a (an) _____.
    (A) vacuum
    (B) negative pressure
    (C) atmospheric pressure
    (D) positive pressure

44. A document, the main text of which contains only mandatory provisions using the word "shall" to indicate requirements, is best described as a _____.
    (A) legal document
    (B) code
    (C) law
    (D) standard

45. A circuit vent connects to a horizontal drain branch and vents _____ traps and not more than _____ traps from a battery of fixtures.
    (A) 2, 6
    (B) 2, 8
    (C) 3, 6
    (D) 3, 9

46. When a backflow prevention device does not bear a critical level marking, the part of the device that shall constitute the critical level is the _____.
    (A) top
    (B) bottom
    (C) center
    (D) outlet invert

47. A water heater is by definition an appliance for _____.
    (A) heating water to a maximum of 210°F
    (B) supplying steam for domestic and commercial purposes
    (C) heating water and equipped with a draft hood
    (D) supplying hot water to a maximum of 190°F

48. The pipe or duct that connects a fuel-gas-burning water heater to a gas vent or chimney is a _____.
    (A) flue collar
    (B) vent connector
    (C) collar connecter
    (D) chimney collar connector

49. A water heater that is constructed and installed so that all air for combustion is derived directly from the outside atmosphere and all flue gases are discharged to the outside atmosphere is a (an) _____.
    (A) direct-vent appliance
    (B) gravity-vent appliance
    (C) forced-draft vent appliance
    (D) induced vent appliance

50. Type B vent is used for venting a listed or approved appliance equipped to burn _____.
    (A) only gas
    (B) only oil
    (C) gas or oil
    (D) solid fuel

51. The piping and equipment between the street gas main and the gas piping system inlet, which is installed by and is under the control and maintenance of the serving gas supplier, is listed in the code as _____.
    (A) gas piping
    (B) service piping
    (C) gas fitting
    (D) yard piping

52. Grounding electrodes are devices that establish an electrical connection to the _____.
    (A) earth
    (B) cold water piping
    (C) building structural steel
    (D) building electrical wiring

53. SCFM means _____.
    (A) standard criteria for medicine
    (B) standard criteria for medical gas
    (C) standard cubic feet per minute
    (D) standard cubic feet per meter

54. A valve that controls the gas or vacuum service to a particular area is a (an) _____.
    (A) zone valve
    (B) branch valve
    (C) area valve
    (D) isolation valve

55. The pipe delivering potable water from the source of supply to a building or structure is the _____.
    (A) water main
    (B) building supply
    (C) water-distributing pipe
    (D) water riser

56. The service valve serves _____.
    (A) vertical piping extending from a horizontal branch
    (B) horizontal piping extending from a riser to a station outlet or inlet
    (C) a designated zone
    (D) all piping downstream from the source

57. The T rating limits the maximum temperature rise of _____ above its initial temperature through the penetration on the nonfire side.
    (A) 400°F
    (B) 300°F
    (C) 325°F
    (D) 375°F

58. The definition of gray water is _____.
    (A) untreated wastewater that has not come into contact with kitchen sink or dishwasher waste
    (B) untreated wastewater that has not come into contact with toilet waste
    (C) wastewater from bathtubs, showers, lavatories, clothes washers and laundry tubs
    (D) all of the above are requirements

59. Reclaimed water is defined as _____.
    (A) untreated domestic wastewater
    (B) tertiary treated nonpotable water
    (C) treated for drinking
    (D) treated rainwater

60. A device located at the bottom of the tank for the purpose of flushing water closets and similar fixtures is a _____.
    (A) ballcock
    (B) tank ball
    (C) flush valve
    (D) flushometer valve

61. Gray water is untreated wastewater that has not come into contact with waste from which one of the following?
    (A) showers
    (B) tubs
    (C) toilets
    (D) lavatories

62. A physical separation that may be a low inlet into the indirect waste receptor from the fixtures, appliances, or devices indirectly connected is the definition for an _____.
    (A) airbreak
    (B) airgap
    (C) indirect waste
    (D) individual vent

63. A pipe that makes an angle of 45 degrees from the vertical is considered _____.
    (A) vertical
    (B) horizontal
    (C) crooked
    (D) offset

64. The definition for the word "stack" does not include _____.
    (A) soil piping
    (B) vent piping
    (C) waste piping
    (D) water piping

65. A medical gas system whose failure can cause minor injury to patients is defined as Category _____.
    (A) 1
    (B) 2
    (C) 3
    (D) 4

66. Air supplied from cylinders, bulk containers or reconstructed from oxygen USP is defined as Medical _____.
    (A) Air
    (B) Vacuum
    (C) Oxygen
    (D) Nitrous Oxide

CHAPTER 3
GENERAL REGULATIONS

This part of the Uniform Plumbing Code is broader in scope, in that it provides the general regulations, instructions, and requirements for the installation and construction of plumbing systems.

It requires that sewage or liquid wastes be disposed of by means of plumbing fixtures connected to a drainage system and terminating into a public or private sewer.

It prohibits depositing any substance into the drainage system that could cause damage to the system. Additional regulations require that design, construction, and workmanship be in conformity with accepted engineering practices. Included are restrictions prohibiting plumbing systems, or parts thereof, from crossing property lines.

The questions in this examination will test you on your knowledge of the aforementioned regulations and also on your understanding of protection of materials and structures, the installation of hangers and supports, and the proper method for trenching, excavating, and backfilling.

1. Disposal of sewage, or other liquid wastes, shall be by means of _____.
   (A) chemical toilets
   (B) sewage-retaining tanks
   (C) the drainage system of the building or premises
   (D) subsurface leaching pits

2. It shall be unlawful for a person to permit the disposal of sewage or other liquid wastes not properly connected to a drainage system.
   (A) True
   (B) False

3. Burred ends of pipe and tubing shall be _____.
   (A) avoided by use of special cutting tools
   (B) confined to exposed and readily accessible locations
   (C) reamed to the full bore of the pipe or tube
   (D) painted to reduce friction

4. Except as otherwise provided in this code, no plumbing system, drainage system, building sewer, private sewage disposal system, or parts thereof, shall be located in _____.
   (A) the walls, floors, or underground in any building
   (B) any lot other than the lot that is the site of the building served by such facilities
   (C) pipe access spaces where accessible for repair
   (D) any garage, or equipment or locker room

5. Piping, fixtures, or equipment that are installed so that they interfere with normal operation and use of windows or doors shall be _____.
   (A) protected from damage
   (B) identified for restricted use
   (C) relocated to avoid interference
   (D) approved if acceptable by the owner

6. All design, construction, and workmanship shall be in accordance with _____.
   (A) accepted engineering practices
   (B) practices of the contractor
   (C) standard shop procedures
   (D) minimizing man hours

7. When supporting horizontal piping 5-inches in diameter with hangers, a ______ inch diameter rod shall be used.
   (A) 3/8
   (B) 1/2
   (C) 5/8
   (D) 1

8. The building drain from the front building may be extended to the rear building when both buildings are located ______ and no private sewer is available or can be constructed.
   (A) on a corner lot
   (B) on an interior lot
   (C) 200 feet from the public sewer
   (D) 300 feet from the public sewer

9. All trenches deeper than the footing of a building or structure, and paralleling the same shall be located not less than ______ from the bottom exterior edge of the footing or as approved.
   (A) 45 degrees
   (B) 90 degrees
   (C) 4 feet
   (D) 12 feet

10. Cast-iron soil pipe, installed vertically, shall be supported at its base and ______.
   (A) each floor not to exceed 15 feet
   (B) every other story
   (C) 5-foot intervals
   (D) 10-foot intervals

11. Threaded or welded 3/4 inch steel pipe supplying water installed horizontally shall be supported at intervals not exceeding ______.
   (A) 5 feet
   (B) 10 feet
   (C) 12 feet
   (D) 15 feet

12. Piping in the ground shall be laid on ______.
   (A) a firm bed
   (B) concrete blocks
   (C) redwood stakes
   (D) common bricks

13. PEX piping may be coupled together with shielded couplings.
   (A) True
   (B) False

14. The type of acceptable joints for PEX-AL-PEX shall be ______.
   (A) metal insert or metal compression
   (B) solvent cemented
   (C) shielded coupling
   (D) mechanical

15. Copper tubing installed vertically shall be supported at each floor or at intervals not to exceed ______.
   (A) 4 feet
   (B) 10 feet
   (C) 8 feet
   (D) 6 feet

16. Mechanical devices such as bulldozers, graders, etc., shall be permitted to complete backfilling to grade where covering above piping is at least ______ inches.
   (A) 6
   (B) 12
   (C) 30
   (D) 36

17. Two-inch copper drain pipe installed horizontally shall be supported at intervals not exceeding ______.
   (A) 4 feet
   (B) 6 feet
   (C) 10 feet
   (D) 12 feet

18. Horizontal hubless pipe shall be braced at not more than 40-foot intervals to prevent _____.
    (A) noise
    (B) leaking
    (C) vertical movement
    (D) horizontal movement

19. All piping, fixtures, and equipment shall be adequately supported in accordance with the manufacturer's installation instructions, the Authority Having Jurisdiction, and the _____.
    (A) contractor
    (B) UPC
    (C) owner
    (D) interior designer

20. Voids around piping passing through concrete floors on the ground shall be _____.
    (A) flanged
    (B) protected
    (C) sealed
    (D) packed

21. All materials, fixtures, or devices used or entering into the construction of plumbing and drainage systems shall be listed as _____.
    (A) complying with approved applicable standards
    (B) conforming to green product standards
    (C) conforming to price of job
    (D) conforming to material availability

22. The provisions of the code allow an alternate material or method of construction provided it is _____.
    (A) approved by the architect as an alternate
    (B) accepted by the owner as an alternate
    (C) approved by the Authority Having Jurisdiction as an alternate
    (D) manufactured by a reputable organization

23. I.P.S. is the abbreviation for _____.
    (A) iron pipe standard
    (B) iron pipe size
    (C) inside pipe standard
    (D) inside pipe size

24. Section 312.2 prohibits the direct embedment of hubless cast-iron pipe in a concrete slab.
    (A) True
    (B) False

25. Tub waste openings in framed construction to crawl spaces at or below the first floor shall be protected by _____.
    (A) completely covering the opening with plywood collars
    (B) the installations of approved metal collars or screens
    (C) flashings
    (D) filling the annular space with approved fiber material

26. It is permitted to drill and tap a vent pipe to connect an indirect waste pipe.
    (A) True
    (B) False

27. All pipe, pipe fittings, traps, fixtures, materials, and devices used in plumbing systems shall be _____.
    (A) listed or labeled by a listing agency
    (B) approved by the Authority Having Jurisdiction
    (C) free from defects
    (D) all of the above

28. Sleeves shall be provided to protect all piping through _____ and _____ walls.
    (A) floors, concrete
    (B) wood, concrete
    (C) concrete, masonry
    (D) none of the above

29. Horizontal hubless cast-iron piping shall be supported at every other joint, unless over ______ feet in length, then it shall be provided with support at every joint.
    (A) 4
    (B) 6
    (C) 8
    (D) 10

30. When installing ______, the Authority Having Jurisdiction shall require evidence of the competency of the installers.
    (A) potable water piping
    (B) fuel-gas piping
    (C) medical gas piping
    (D) plumbing appurtenance

31. Drain pipes subject to condensation over food storage in food-handling establishments shall be ______.
    (A) installed with a spill pan
    (B) thermally insulated
    (C) rerouted out of the area
    (D) tested with 5 psi of air

32. Test gauges shall have a pressure range not exceeding ______.
    (A) the test pressure applied
    (B) twice the test pressure applied
    (C) 15 psi
    (D) 60 psi

33. If an Authority Having Jurisdiction permits an "alternative engineered design," ______.
    (A) the approval shall have effect beyond that Authority Having Jurisdiction boundary
    (B) the approval shall have no effect beyond that Authority Having Jurisdiction boundary
    (C) a property owner shall also approve the change
    (D) a covenant shall be recorded to that effect

34. Plumbing systems shall be located above the design flood elevation ______.
    (A) under no circumstances
    (B) unless approved as a buoyancy compliant system
    (C) except when designed to prevent water from entering or accumulating within their components
    (D) if they contain backwater valves

35. Sound limitations in plumbing systems shall be designed and installed in accordance with the ______ code.
    (A) mechanical
    (B) fire
    (C) energy
    (D) building

36. One hundred feet of horizontal hubless cast iron pipe include two vertical stack connections. Per Table 313.3 how many hangers will be required on the horizontal run using 10-foot lengths of hubless cast iron pipe?
    (A) 9
    (B) 10
    (C) 18
    (D) 20

CHAPTER 4

PLUMBING FIXTURES AND FIXTURE FITTINGS

As the title suggests, this part of the Uniform Plumbing Code contains regulations relative to plumbing fixtures. It controls the quality of the fixture and has provisions allowing the construction of special fixtures for certain establishments.

Regulations are given for connections to fixtures, the material requirement for strainers, strainer bodies, lock nuts, sizes of continuous wastes, the need for directional tees, and access panels. Included are regulations for securing fixtures to the walls or floors, and also for setting fixtures with respect to clearances from side walls and partitions. Detailed requirements are given for the construction, installation and testing of shower receptors.

Current trends in the plumbing industry show that field installations of plumbing fixtures are becoming increasingly popular. The important observation in this respect is to ensure that the fixture fittings are not installed so as to compromise the designated mounting surface, and comply with backflow requirements.

1. Screws or bolts for securing floor-mounted fixtures shall be _____.
   (A) copper alloy
   (B) aluminum
   (C) galvanized steel
   (D) die-cast zinc

2. Fixtures having concealed slip-joint connections shall be provided with _____.
   (A) a utility space or access panel
   (B) a ladder and a light fixture
   (C) red brass fittings
   (D) rubber gaskets

3. The overflow pipe from a lavatory shall be connected on the _____.
   (A) house or inlet side of the fixture trap
   (B) sewer or outlet side of the fixture trap
   (C) downstream side of the fixture tee
   (D) upstream side of the fixture tee

4. The diameter of a waste outlet and fixture tailpiece for a bathtub shall not be less than _____.
   (A) 1-1/4 inches in diameter
   (B) 1-3/8 inches in diameter
   (C) 1-1/2 inches in diameter
   (D) 1-3/8 inches internal diameter

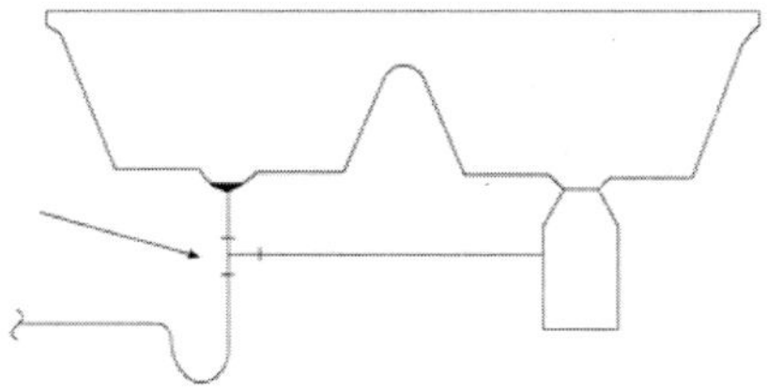

Figure 5

5. The fitting indicated by the arrow shall be a(n) _____.
   (A) sanitary tee
   (B) directional-type branch
   (C) approved wye
   (D) B & C

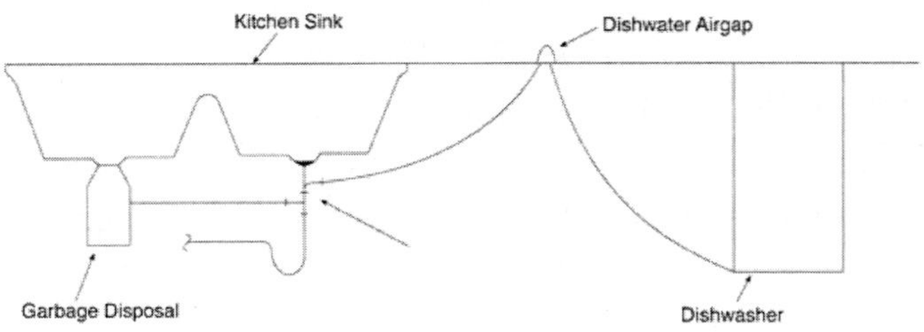

Figure 6

6. The connection of the dishwasher drain directly to the continuous waste, as indicated by the arrow, is _____.
   (A) a code violation
   (B) good practice
   (C) not a code violation
   (D) a cross-connection

7. No dry or chemical closet (toilet) shall be installed in any building used for human habitation unless first approved by the _____.
   (A) owner
   (B) plumbing inspector
   (C) city clerk
   (D) health officer

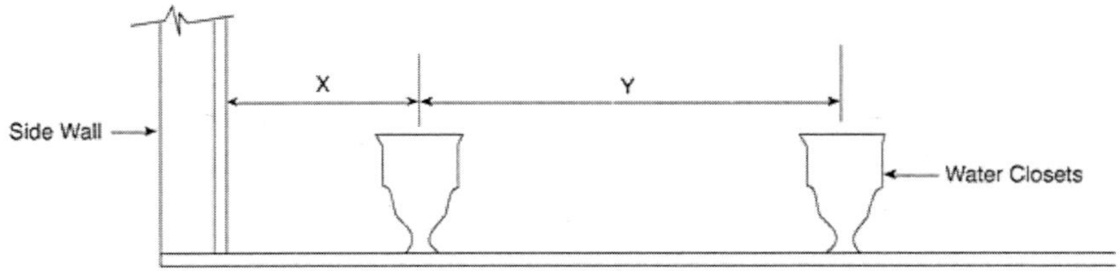

Figure 8

8. The distance X shall be not less than _____.
   (A) 12 inches
   (B) 15 inches
   (C) 24 inches
   (D) 30 inches

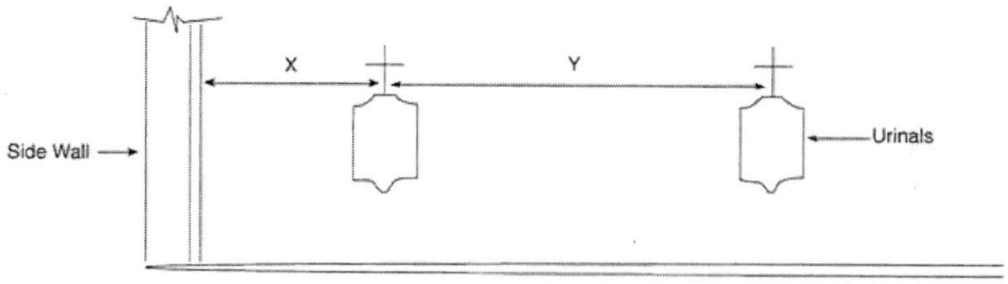

Figure 9

9. The distance Y shall be not less than _____.
   (A) 12 inches
   (B) 15 inches
   (C) 24 inches
   (D) 30 inches

10. Tanks flushing more than one urinal shall be _____.
   (A) automatic in operation
   (B) continual in operation
   (C) manual in operation
   (D) acid-resistant enamel

11. A water supply to a urinal shall be protected by an approved _____.
    (A) relief valve
    (B) vacuum breaker
    (C) metering valve
    (D) automatic switch

12. Trough urinals shall be _____.
    (A) permissible if constructed of concrete
    (B) required in large associations
    (C) prohibited
    (D) allowed in bus stations

13. With two noted exceptions, all shower compartments and receptors shall be capable of encompassing a _____ diameter circle.
    (A) 24 inch
    (B) 30 inch
    (C) 33 inch
    (D) 36 inch

14. The curb or threshold of a shower compartment is required for _____.
    (A) installations in accordance with ICC A117.1 and accessibility standards
    (B) gang showers
    (C) shower receptors built on-site
    (D) none of the above

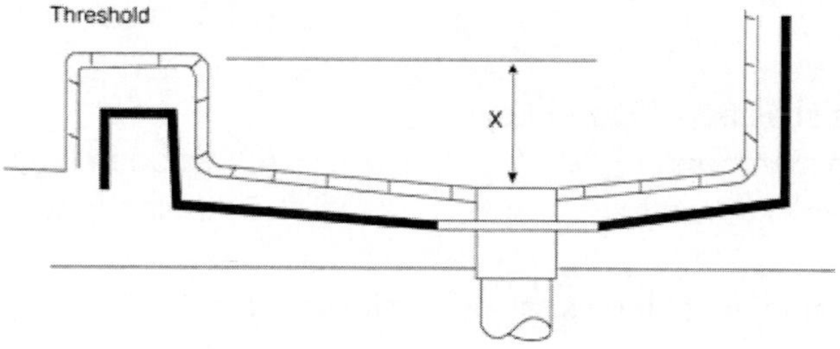

Figure 15

15. If provided with a threshold or curb, distance X shall not be less than _____ or exceed _____.
    (A) 2 inches, 9 inches
    (B) 1 inch, 7 inches
    (C) 3 inches, 10 inches
    (D) 2 inches, 7 inches

16. Shower pan linings may be constructed with _____.
    (A) 26 B&S gauge copper
    (B) 3 pounds per square foot lead
    (C) 15 pounds asphalt felt
    (D) zinc-coated steel

17. Joints in copper shower pan liners shall be _____.
    (A) burned
    (B) swaged
    (C) sealed with epoxy cement
    (D) soldered or brazed

18. Drinking fountains shall not be installed in _____.
    (A) public toilet rooms
    (B) restaurants
    (C) bars
    (D) schoolyards

19. Water closets set side by side shall be no closer than _____.
    (A) 15 inches center-to-center
    (B) 18 inches center-to-center
    (C) 24 inches center-to-center
    (D) 30 inches center-to-center

20. The minimum distance from a side wall or partition to the center of a urinal shall be ______ inches.
    (A) 6
    (B) 12
    (C) 18
    (D) 24

21. Manually controlled flushometer valves shall be used to flush not more than ______.
    (A) one urinal
    (B) two urinals
    (C) three urinals
    (D) four urinals

22. The joint of a wall-hung lavatory, where it is in contact with the wall, shall be ______.
    (A) smooth
    (B) painted
    (C) concealed
    (D) made watertight

23. Approved type bonderized and galvanized sheet steel of not less than 16 U.S. gauge may be used to make dishwashing sinks for ______.
    (A) restaurants
    (B) lunch counters
    (C) commercial buildings
    (D) residential buildings

24. Gutters in public shower rooms shall be sloped not less than ______.
    (A) 5 percent
    (B) 10 percent
    (C) 2 percent
    (D) 3 percent

25. Wall-hung lavatories shall be supported by ______.
    (A) metal supporting members
    (B) rigid traps
    (C) wooden legs
    (D) their connections

26. The tailpiece diameter for lavatories shall not be less than ______.
    (A) 1-1/4 inches
    (B) 1-3/8 inches
    (C) 1-1/2 inches
    (D) 1-3/4 inches

27. In the construction of "on-site built-up shower receptors", watertight shower pan linings shall be installed, over a smooth and solid sub-base, with a pitch to the "weep holes" of ______.
    (A) 1/8 inch per foot
    (B) 1/4 inch per foot
    (C) 1/16 inch per foot
    (D) 1/32 inch per foot

28. Strainers having waterway areas at least equivalent to the area of their tailpiece are required for ______.
    (A) shower drains
    (B) floor drains
    (C) urinals
    (D) A & B

29. Showers and tub-shower combinations shall be provided with individual control valves of the ______.
    (A) pressure balance type
    (B) thermostatic mixing valve type
    (C) 140°F limit control type
    (D) A and B apply

30. The clear space in front of any water closet, lavatory or bidet shall not be less than ______.
    (A) 12 inches
    (B) 18 inches
    (C) 24 inches
    (D) 30 inches

31. Non-water urinals ______.
    (A) are prohibited by code
    (B) are exempt from plumbing product listings
    (C) require backwater valves
    (D) require individually roughed-in water lines

32. Maximum hot water temperature discharging from a bidet shall be limited to _____.
   (A) 105°F
   (B) 110°F
   (C) 120°F
   (D) not limited

33. Closet bends or stubs shall be cut off so as to present a smooth surface, even with the _____.
   (A) top of the rough floor
   (B) top of the closet ring
   (C) top of the finish floor
   (D) bottom of the closet

34. Wall-mounted water closets shall be securely bolted to _____.
   (A) the wall
   (B) double two by six studs
   (C) a carrier fitting
   (D) a closet bend

35. Fixture connections between drainage pipes and water closets shall be made with approved _____.
   (A) expansion joints
   (B) unions
   (C) flanges
   (D) hot-poured compound

36. The water supply to a bathtub and whirlpool bathtub filler valve shall be protected by an _____ or in accordance with Section 417.0.
   (A) air break
   (B) air gap
   (C) ASSE 1017
   (D) ASSE 1070

37. A domestic dishwashing machine shall discharge indirectly through an air gap fitting into a _____.
   (A) wye branch fitting on the tailpiece of a kitchen sink
   (B) waste receptor
   (C) dishwasher connection of a food waste disposer
   (D) all of the above

38. The path of travel to an emergency eyewash shall be _____.
   (A) clearly identified
   (B) in a straight line
   (C) no more than 50 feet
   (D) no more than 100 feet

39. Supports for off the floor public use plumbing fixtures are to comply with ASME _____.
   (A) A112.6.1.M
   (B) A112.6.2
   (C) A112.19.3
   (D) A112.19.19

40. An emergency eye wash supplied with hot and cold water shall be controlled by a temperature actuated mixing valve complying with ASSE _____.
   (A) 1016
   (B) 1069
   (C) 1070
   (D) 1071

For question 41 the following scenario applies:

A new public school, with an intended use for 1800 students and staff, is to be designed with male, female, and six all gender restrooms.

41. What is the total number of water closets required for the building?
   (A) 28
   (B) 38
   (C) 48
   (D) 58

42. All gender restrooms shall include a urinal in addition to other fixtures.
   (A) True
   (B) False

CHAPTER 5
WATER HEATERS

The regulations of this chapter govern the construction, location, and installation of fuel-burning and other (i.e., electric) water heaters heating potable water, together with all chimneys, vents, and their connectors.

This chapter pertains to listed and/or approved potable water heaters with a nominal capacity of 120 gallons (454L) or less, having a heat input of 200,000 Btu/h (58.62 kW) or less, at a pressure of 160 psi (1,103 kPa) or less, and at temperatures not exceeding 210°F (99°C). See UMC, Section 1001.0, Exception 1.

Any water heater exceeding the above criteria is a hot water-heating boiler regulated by the Uniform Mechanical Code, Chapters 9 and 10. See UMC, definition for Hot-Water-Heating Boiler.

Gas-fired water heaters requiring vents need provisions for combustion air, required clearances for servicing, and clearance to combustible constructions. They are not permitted in any bedroom or bathroom unless direct-vent type or installed in a closet that is equipped with a listed, gasketed door assembly and a listed self-closing device. All combustion air must be obtained from the outdoor. None of the foregoing requirements applies to electric water heaters.

In the past, the approval agency for gas-fired water heaters was the American Gas Association (AGA), but in 1997 AGA transferred its ownership interest in the product testing and certification program to the Canadian Standards Association (CSA).

The requirements for temperature and pressure (T&P) relief valves and pressure relief valves must be strictly adhered to since they provide safety measures against overpressure and overheating. The potential explosion hazard exists when improperly installed.

1. It shall be unlawful for any person to install, remove, or replace any water heater without first _____.
   (A) obtaining a plumbing contractor license from the Authority Having Jurisdiction
   (B) obtaining a permit from the Authority Having Jurisdiction
   (C) having a plumbing journeyman's license
   (D) having permission from the building owner

2. Vent connectors of natural draft appliances _____ be connected to mechanical draft systems operating under positive pressure.
   (A) may
   (B) shall not
   (C) must
   (D) must be interlocked before they can

3. A vent connector connected to a chimney flue serving a fireplace is _____.
   (A) prohibited unless the fireplace flue opening is permanently sealed
   (B) prohibited
   (C) required to be a direct-vent type
   (D) required to use Type B connector

4. A draft hood-equipped water heater is vented through an unconditioned area. The portion of the vent passing through the unconditioned area shall _____.
   (A) not exceed 6 feet in length
   (B) be a minimum 5 inches in diameter
   (C) be listed Type B, Type L, or listed material having equivalent insulating qualities
   (D) be a minimum 24 gauge material

5. The minimum capacity for a storage water heater is ______.
   (A) not regulated by code
   (B) required to be a minimum of 40 gallons
   (C) required to be a minimum of 50 gallons
   (D) based on the "first hour rating" and the number of bedrooms and bathrooms

6. A water heater installed in a residential garage shall be ______.
   (A) listed for a garage installation
   (B) located or protected from damage by a moving vehicle
   (C) located not less than 16 inches above the garage floor
   (D) a direct-vent water heater

7. A final water heater inspection shall be made after ______.
   (A) all plumbing fixtures have been installed
   (B) all gas appliances have been installed
   (C) work authorized by the permit has been installed
   (D) the plumber has left the job

8. Water heaters of other than the direct-vent type shall be located ______.
   (A) within 6 feet of the chimney or gas vent
   (B) within 8 feet of the chimney or gas vent
   (C) as close as practical to the chimney or gas vent
   (D) within 10 feet of the chimney or gas vent

9. A three-bedroom home with two bathrooms is equipped with a gas storage-type water heater. The minimum capacity for the water heater in accordance with the first hour ratings is ______ gallons.
   (A) 38
   (B) 49
   (C) 62
   (D) 74

10. Outdoor combustion air shall be supplied by ______.
    (A) 1-1/2 inch Schedule 80 PVC
    (B) 2 inch Schedule 40 PVC
    (C) one or two permanent openings
    (D) 18 gauge round duct

11. Indoor combustion air supplied by using the "Standard Method" requires a minimum volume of ______.
    (A) 100 cu. ft. per 1,000 Btu/hr
    (B) 50 cu. ft. per 1,000 Btu/hr
    (C) 50 cu. ft. per 2,000 Btu/hr
    (D) 25 cu. ft. per 2,000 Btu/hr

12. Combustion air ducts shall ______.
    (A) serve a single space
    (B) be equal to the size of the water heater vent
    (C) be a minimum of 12 square inches
    (D) be located within the upper or lower 12 inches of the water heater

13. Combustion air ducts that terminate in an attic shall _____.
    (A) be screened at the attic end only
    (B) be screened at both ends
    (C) not be screened
    (D) not be screened unless screen openings are at least 1 inch minimum

14. Draft hood-equipped and other fuel-burning Category I water heaters shall be permitted to be installed in a closet located in a bedroom or a bathroom provided _____.
    (A) the appliance is listed
    (B) the closet is of fire-resistant construction
    (C) the closet has a listed, gasketed door assembly and a listed self-closing device
    (D) combustion air is provided by adjacent room

15. Water heaters supported from the earth shall rest on _____.
    (A) listed concrete blocks
    (B) a level concrete or other approved base
    (C) level solid undisturbed earth
    (D) a listed base of noncombustible construction

16. Where damage may result from a leaking water heater, a watertight pan of corrosion-resistant materials shall be installed beneath the water heater when it is located in an attic or _____.
    (A) second-floor mechanical room
    (B) garage
    (C) any story above the first story
    (D) a floor-ceiling assembly

17. Plastic piping may be used to vent a water heater if the water heater is _____.
    (A) a noncondensing type
    (B) listed for use with that material
    (C) a direct-vent type
    (D) an induced-or force-draft type

18. Special gas vents shall only be installed on a water heater if _____.
    (A) a gravity vent is not practicable
    (B) head height above the water heater is less than 1 foot
    (C) it is listed and installed per the listing and the manufacturer's instructions
    (D) the fire department or fire marshal deems it to be necessary

19. The minimum horizontal distance that a category IV vent terminates through an outside wall of a building shall be ____ feet from an operable opening in an adjacent building.
    (A) 7
    (B) 10
    (C) 12
    (D) 15

20. Water heater draft hoods or barometric draft regulators shall be _____.
    (A) of the same size as the vent connector
    (B) located outside the appliance enclosure
    (C) located as close as practicable to the water heater
    (D) installed in the same room or enclosure as the water heater

21. The maximum horizontal length of a 5-inch vent connector for a water heater shall not exceed _____.
    (A) 3 feet
    (B) 6 feet
    (C) 7-1/2 feet
    (D) 4 feet

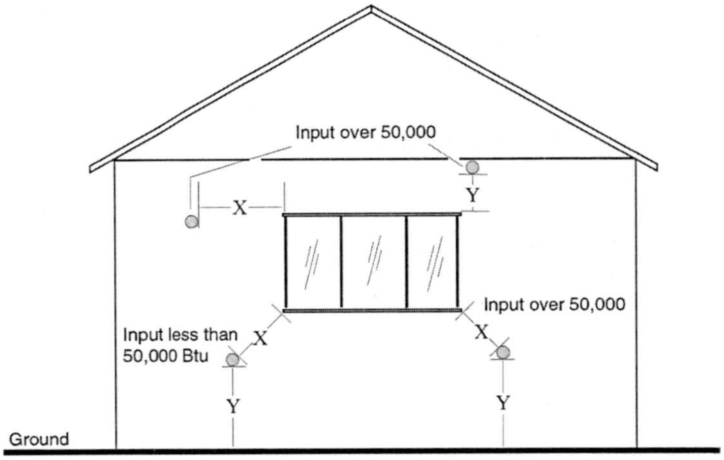

22. Referring to the above drawing for a direct vent appliance with an input of 50,000 Btu or less, the minimum distance at X shall be _____ inches. The minimum distance at Y shall be at least _____ inches.
(A) 12, 9
(B) 12, 12
(C) 9, 12
(D) 9, 9

23. Referring to the above drawing for a direct vent appliance with an input greater than 50,000 Btu, the minimum distance from an opening into a building shall be _____ inches. The minimum distance of the bottom of the vent terminal above finish grade shall be _____ inches.
(A) 12, 9
(B) 12, 12
(C) 9, 12
(D) 9, 9

24. A Type B vent with a listed cap of 14 inches shall terminate at least _____ higher than any portion of a building within 10 feet.
(A) 1 foot
(B) 2 feet
(C) 18 inches
(D) 3 feet

25. A water heater is installed on the roof of a building more than 15 feet in height. The code requires _____ to the roof unless other means acceptable to the Authority Having Jurisdiction are used.
(A) a permanent ladder outside the building
(B) a portable ladder
(C) an outside means of access
(D) an inside means of access

26. When using an alternative method of sizing with two vent connectors for draft hood-equipped water heaters, the effective area of the common vent connector or vent manifold and all junction fittings shall not be less than the area of the larger vent connector plus _____ percent of the areas of smaller flue collar outlets.
(A) 50
(B) 75
(C) 25
(D) 10

27. Water heaters shall be anchored or strapped to resist horizontal displacement due to earthquake motion.
(A) True
(B) False

28. Where utilizing louvers and grills to provide combustion air, and the design and free area are not known, it shall be assumed that wood louvers will have _____ percent free area and metal louvers and grills will have _____ percent free area.
(A) 50, 25
(B) 25, 75
(C) 25, 50
(D) 50, 75

29. Balancing baffles shall be _____.
    (A) mechanically locked
    (B) electronically controlled
    (C) pneumatically controlled
    (D) 80 percent open

30. Clearances for listed water heaters shall be _____.
    (A) a minimum of 1 inch to combustible construction
    (B) a minimum of 2 inches to combustible construction
    (C) in accordance with their listing and the manufacturer's instructions
    (D) a minimum of 6 inches to unprotected combustible construction

31. Water heaters installed in residential garages shall be installed so that all burners and burner-ignition devices are located not less than 18 inches above the floor unless _____.
    (A) approved by the fire authority
    (B) the water heater has a separate combustion chamber
    (C) the water heater is listed as flammable vapor ignition resistant
    (D) the garage is detached from the living spaces

32. Water heaters not listed for outside installation but installed outdoors shall be provided with _____.
    (A) protection to the degree that the environment requires
    (B) smoke and antitheft alarms
    (C) a 1-hour fire-rated roof
    (D) a listed anticorrosion coating

33. A Type B gas vent shall terminate at least _____ feet above the highest connected equipment draft hood or flue collar.
    (A) 12
    (B) 6
    (C) 5
    (D) 10

34. Garage appliances may be installed in a separate enclosed space having access only from the outside providing the combustion air is _____.
    (A) provided from the enclosure
    (B) supplied from the garage
    (C) taken from the exterior of the garage
    (D) supplied from an attic

35. Indirect-fired water heaters that incorporate a single-wall heat exchanger shall _____.
    (A) have a maximum operating pressure in the heat exchanger not exceeding the maximum operating pressure of the potable water supply
    (B) have a heat-transfer medium of potable water or other nontoxic fluid recognized as food grade by the FDA
    (C) bear the proper "caution" labels
    (D) meet all of the above

36. A listed water heater, located in an interior floor level closet, must have an access dimension _____ for servicing purposes.
    (A) 24"w x 30"h
    (B) in accordance with their listing and the manufactures instructions
    (C) 30"w x 36"h
    (D) the size of the water heater

37. The passageway for a water heater in an attic shall be unobstructed and shall have a solid flooring not less than _____ inches wide from the entrance opening to the water heater.
    (A) 22
    (B) 30
    (C) 24
    (D) 36

38. For attic installation, the passageway and servicing area adjacent to the appliance shall be _____.
    (A) identified
    (B) well lighted
    (C) dry
    (D) floored

39. Above ceiling venting systems or non-ducted air handling systems passing through an above ceiling air space or other non-ducted portion shall meet one of the following:
    (A) shall be a listed special gas vent
    (B) fittings and joints shall be installed above ceiling
    (C) no conduits or enclosures are required
    (D) joints shall be airtight

40. When venting an appliance with plastic pipe, it is important _____.
    (A) not to use transition fittings
    (B) to use no more than 4 elbows in a 20' run of piping
    (C) that vent piping shall be graded at a minimum of 2% slope
    (D) to identify the piping material by the manufacturer's instructions

41. When gas vents are installed in a masonry chimney, _____.
    (A) a permanent label is needed where the vent enters the chimney
    (B) it must be installed in accordance with the Authority Having Jurisdiction
    (C) the piping must be accessible for maintenance and cleaning
    (D) the vent material is needs to be airtight

42. For each additional elbow up to and including 45 degrees, the maximum vent connector capacity listed in the venting tables, where two 90 degree elbows are used, shall be reduced by what percentage?
    (A) 0.5
    (B) 5
    (C) 6.25
    (D) 50

CHAPTER 6
WATER SUPPLY AND DISTRIBUTION

Each plumbing fixture shall be provided with an adequate supply of potable running water, piped thereto in an approved manner, so arranged as to flush and keep it in a clean and sanitary condition without danger of backflow or cross-connection.

The design and installation requirements for water piping systems are technical engineering subjects that have been simplified by the Uniform Plumbing Code, so that they can be easily understood. Regulations are given for the type of material approved for water piping and the acceptable locations, including requirements for shut-off valves, water pressure regulators, and pressure-relief valves.

The procedure for sizing water piping systems is given for either the method using established tables or an additional method used for larger systems. The latter method involves establishing tables by means of compiled data and calculations.

Prevention of contamination or pollution of the potable water supply system is most important for obvious reasons. The primary requirement is for an "airgap" between a water outlet and the fixture overflow level. Where this is not practical, backflow prevention devices are required and approved for the potential hazard. There are several types of backflow prevention devices. Each is used for a specific purpose and each has a specific method of installation.

1. The code requirement for "an adequate supply of potable running water" is for _____.
   (A) cleansing and sanitation
   (B) protection of trap seals
   (C) water-cooled equipment
   (D) hydronic systems

2. The primary reason for installing backflow prevention assemblies is to _____.
   (A) prevent excessive pressure
   (B) prevent mixing of hot and cold water
   (C) prevent pollution or contamination of the potable water supply
   (D) prevent scalding

3. Ballcocks in water closet tanks shall be installed so that the critical level of the water inlet is above the full opening of the overflow pipe of not less than _____.
   (A) 1/2 inch
   (B) 1 inch
   (C) 3 inches
   (D) 6 inches

For questions 4 through 6, choose the answer you think is correct for the minimum height of the atmospheric vacuum breaker critical level above the point specified.

4. Above the highest part of a urinal equipped with a flushometer valve _____.
   (A) 1 inch
   (B) 3 inches
   (C) 6 inches
   (D) 12 inches

5. Above the overflow rim of a tank, vat or swimming pool _____.
   (A) 1 inch
   (B) 2 inches
   (C) 6 inches
   (D) 12 inches

6. Above all downstream piping for a lawn sprinkler system _____.
   (A) 1 inch
   (B) 6 inches
   (C) 12 inches
   (D) 24 inches

7. PEX tubing shall not be installed within the first _____ of piping connected to a water heater.
   (A) 12 inches
   (B) 18 inches
   (C) 24 inches
   (D) 36 inches

8. Each outlet on a nonpotable water system that is used for special purposes shall be posted _____.
   (A) CAUTION – PROCESSED WATER
   (B) CAUTION – CONTAMINATED WATER
   (C) CAUTION – POLLUTED WATER
   (D) CAUTION – NONPOTABLE WATER DO NOT DRINK

9. The piping material(s) not approved in the cold water piping located inside a building is _____.
   (A) PVC
   (B) PE
   (C) PE-RT
   (D) A & B

10. All malleable iron water fittings shall be _____.
   (A) 2 inch or smaller
   (B) galvanized
   (C) flanged
   (D) removed

11. Water valves 2-1/2 inches and larger may have bodies that are _____.
   (A) galvanized
   (B) die-cast
   (C) aluminum core
   (D) cast iron

12. Changes in direction of copper tubing may be made with _____.
   (A) non deforming bending equipment
   (B) heat annealing
   (C) ferrous fittings
   (D) plastic fittings

13. Water piping systems shall be designed on the basis of the _____.
   (A) maximum pressure
   (B) minimum pressure
   (C) average residual pressure
   (D) average service pressure

14. Copper water tube Type "M" may be installed within a building _____.
   (A) below slab only
   (B) above ground only
   (C) for cold water only
   (D) if it is wrapped

15. Immediately ahead of each water-supplied appliance and each slip joint or appliance supply a _____ is required.
   (A) control valve
   (B) check valve
   (C) backwater valve
   (D) solenoid valve

16. After allowing for friction and other pressure losses, the residual water pressure shall be at least _____.
   (A) 0.433 psi
   (B) 7.5 psi
   (C) 15 psi
   (D) 29 psi

17. A water piping system with a pressure of 110 psi will require a (an) _____.
   (A) reduced pressure backflow device
   (B) extra-heavy pipe and fittings
   (C) pressure regulator and strainer
   (D) pressure-relief valve

18. Any water heater connected to a separate storage tank and having valves between the heater and tank shall be provided with a _____.
    (A) fullway valve
    (B) fullway gate valve
    (C) water pressure regulator
    (D) water pressure-relief valve

19. Combination temperature and pressure relief valves shall be installed _____.
    (A) based on their listing requirements and the manufacturer's instructions
    (B) within 3 inches to 6 inches of the water heater
    (C) in lieu of pressure-relief valves
    (D) on all nonstorage-type water heaters

20. Galvanized steel piping installed under a concrete floor slab within a building shall _____.
    (A) have a machine-applied coating
    (B) have a spiral wrap
    (C) be installed without joints
    (D) be installed with brazed joints

21. Cast-iron fittings, up to and including two inches in size, when used in connection with potable water piping, shall be _____.
    (A) extra heavy
    (B) flanged
    (C) disallowed
    (D) galvanized

22. Water piping may be installed in the same trench with less than 12 inches of separation from a building sewer constructed of _____.
    (A) cast iron
    (B) clay
    (C) asbestos cement
    (D) bituminized fiber

23. The available pressure at the water meter is 80 psi. At the highest supply outlet, 36 feet above the water meter, the pressure would be _____ psi when sizing a water system according to Chapter 6.
    (A) 62
    (B) 44
    (C) 116
    (D) 80

24. Building supplies shall be no smaller than that required for the installation and, in no case, smaller than _____ inch(es).
    (A) 1/2
    (B) 3/4
    (C) 1
    (D) 2

25. Copper tubing used for water piping under a concrete slab within a building shall be _____.
    (A) installed with brazed joints with cast brass fittings
    (B) protected with a factory-applied coating
    (C) installed without joints where possible
    (D) installed with Type K copper tube with brazed joints

26. CPVC water pipe may be used for hot and cold water distribution systems _____.
    (A) in residences only
    (B) within a building
    (C) belowground only
    (D) aboveground only

27. Copper or stainless steel flexible connectors for water heater connectors shall be limited to _____.
    (A) 12 inches
    (B) 18 inches
    (C) 24 inches
    (D) 36 inches

28. Unions shall be installed in a water supply system within 12 inches of _____.
    (A) water heating tanks
    (B) conditioning tanks
    (C) regulating equipment
    (D) all of the above

29. The type of valve required on a building supply to any building is a _____.
    (A) fullway valve
    (B) corporation valve
    (C) butterfly valve
    (D) globe valve

30. All backflow devices or assemblies shall be tested by a _____.
    (A) licensed agent
    (B) representative from the Authority Having Jurisdiction
    (C) certified backflow assembly tester
    (D) third-party certified agency

31. Other than systems sized by the use of Table 610.4, the maximum velocity of a cold water system piped with copper and copper alloy tube and fittings shall be limited to _____.
    (A) 5 feet per second
    (B) 8 feet per second
    (C) 10 feet per second
    (D) 15 feet per second

32. Discharge piping serving a pressure-relief shall be provided with a _____.
    (A) full size drain
    (B) check valve to prevent backflow
    (C) trapped drain line
    (D) drain that terminates 30 inches or more above grade

33. The quantity of water required to be supplied to every plumbing fixture shall be represented by _____.
    (A) gallons per minute
    (B) feet per second
    (C) fixture units
    (D) available head pressure

34. The size and material of irrigation water piping installed outside of any building and separated from the potable water supply by means of an approved backflow prevention device is _____.
    (A) regulated by code
    (B) not regulated by code
    (C) regulated by the National Sanitation Foundation
    (D) regulated by the American Society of Sanitary Engineers

35. Hard-drawn copper tubing, marked with a blue stripe, is referred to as _____.
    (A) Type "K"
    (B) Type "L"
    (C) Type "M"
    (D) Type "DWV"

36. The color stripe on Type "K" hard-drawn copper tubing is _____.
    (A) green
    (B) blue
    (C) red
    (D) yellow

37. Flaring tools are used to make joints for _____.
    (A) hard Type K copper tubing
    (B) soft copper water tubing
    (C) yellow brass water piping
    (D) aluminum water piping

38. The distance between a water heater and the union on the water lines shall not be more than _____ inches.
(A) 6 (C) 18
(B) 12 (D) 24

39. Whenever fixtures and/or fixture fittings are installed that require a residual water pressure higher than 15 pounds per square inch _____.
(A) that minimum pressure shall be provided
(B) provisions shall be made to use fixtures that do not require higher residual pressure
(C) the piping system shall be not less than copper tubing Type L
(D) A and C apply

40. _____ water piping systems shall be permitted to be tested with air.
(A) PVC (C) CPVC
(B) PEX (D) PB

41. The duration for a water system pressure test shall be for a period of not less than _____ minutes.
(A) 5 (C) 15
(B) 10 (D) 30

42. In occupancies where plumbing fixtures are installed for _____, hot water shall be required for bathing, washing, laundry, cooking and similar purposes.
(A) public use (C) private use
(B) commercial use (D) daycare facilities

43. Except as required in Section 601.3.3, nonpotable water systems shall have a _____ with black uppercase lettering, with the words "CAUTION: NONPOTABLE WATER, DO NOT DRINK."
(A) white background (C) green background
(B) blue background (D) yellow background

44. Plastic materials for _____ piping outside underground shall have an electrically continuous corrosion-resistant blue insulated copper tracer wire or other approved conductor installed adjacent to the piping.
(A) building supply (C) irrigation
(B) well (D) hose bibb

45. A water distribution system with a device or arranged in a way that prevents dissipation of building pressure back into the water main shall be provided with an approved, listed, and adequately sized _____ or other approved device.
(A) double-check valve (C) temperature & pressure relief valve
(B) expansion tank (D) PRV

46. Pipe joint material shall be insoluble in water, nontoxic, and applied _____.
(A) on male threads only (C) on inside fitting threads only
(B) on pipe threads and inside fittings (D) in warm weather only

47. Surfaces to be joined by soldering or brazing shall be cleaned bright by _____.
(A) mechanical or manual means (C) a corrosive flux
(B) a self-cleaning flux (D) sulfuric acid

48. Solder and fluxes used in piping systems conveying potable water are prohibited when the lead content exceeds ______ percent.
    (A) 0.2
    (B) 0.5
    (C) 1
    (D) 1.5

49. The minimum wall thickness for threaded plastic pipe (not molded fittings) shall be ______.
    (A) Schedule 40
    (B) Schedule 80
    (C) 1/8 inch
    (D) 1/16 inch

50. When brazing, the alloys shall have a liquid temperature above ______ degrees F.
    (A) 800
    (B) 840
    (C) 1,000
    (D) 1,200

51. A push-fit fitting is ______.
    (A) assembled using excessive force
    (B) assembled by pushing tube/pipe into the fitting and is sealed with an "O" ring that forms the joint
    (C) prohibited by Code
    (D) permitted to be installed in accessible locations only

52. Heat-fusion joints ______.
    (A) are used in some thermoplastic systems
    (B) are low-temperature, metallic pipe welds
    (C) require prior approval of an Authority Having Jurisdiction
    (D) may be used in drainage systems but not pressure systems

53. ______ systems shall be clearly identified, have a purple background with uppercase yellow lettering.
    (A) Gray water
    (B) Reclaimed (recycled) water
    (C) Rainwater catchment
    (D) None of the above

54. The area of coverage of a single fire sprinkler shall not exceed ______.
    (A) 100 square feet
    (B) 200 square feet
    (C) 300 square feet
    (D) 400 square feet

55. What size drain for the fire sprinkler system shall be provided on the system side of the water distribution shut off valve?
    (A) 1/2 inch
    (B) 3/4 inch
    (C) 1 inch
    (D) 1-1/2 inch

56. Non metallic pipe and tubing for fire sprinklers shall be protected from exposure to the occupied space by a material having a minimum fire rating of ______ minutes?
    (A) 15
    (B) 30
    (C) 45
    (D) 60

57. In locations where intermediate temperature sprinklers are required, the minimum distance from a kitchen range top (heat source) is ______.
    (A) 6 inches
    (B) 9 inches
    (C) 18 inches
    (D) 24 inches

58. The minimum separation from a wood burning stove to a fire sprinkler shall be measured in a straight line from the nearest edge of the sprinkler to _____.
(A) the centerline of the wood burning stove
(B) the nearest edge of the wood burning stove
(C) the far edge of the wood burning stove
(D) the nearest igniter

59. The sprinkler pipe size from the water supply source to a sprinkler shall not be less than _____ inch.
(A) 3/8
(B) 1/2
(C) 3/4
(D) 1

60. What is the minimum temperature rating of intermediate temperature fire sprinklers when installed in attics and concealed spaces directly beneath a roof?
(A) 135°F
(B) 170°F
(C) 175°F
(D) 225°F

61. The water distribution size of the fire sprinkler system in a single-family home has been determined to be 1 inch. The available pressure at the service is 55 psi. The water meter listing specifies a flow rate of 14 gpm. With a water service developed length of 38 feet, the available psi for fire sprinklers is _____.
(A) 49.8
(B) 50.8
(C) 51.8
(D) 55.0

62. The minimum water test time for a fire sprinkler system is _____ minutes.
(A) 15
(B) 30
(C) 60
(D) 120

63. Nonmetallic fire sprinkler pipe shall be certified for residential sprinkler installations and shall have a pressure rating of not less than _____ psi at 120°F.
(A) 100
(B) 120
(C) 130
(D) 150

64. The water supply for a one story dwelling unit of 1785 square feet shall have the capacity to provide the required flow rate to the sprinklers for a period _____ minutes.
(A) 5
(B) 7
(C) 10
(D) 15

65. Which of the following is an approved method of protecting the water supply for a chemical dispenser?
(A) Double check assembly
(B) Air gap
(C) Check valve
(D) Not needed

66. A _____ joint is an example of a mechanical joint for CPVC Plastic pipe.
(A) welded
(B) soldered
(C) solvent welded
(D) flanged

67. Fittings that employ a quick assembly push fit connector shall comply with _____.
(A) ASSE 1053
(B) ASSE 1061
(C) ASSE 1069
(D) ASSE 1079

68. What is the total demand in fixture units and minimum pipe size for a branch serving a domestic kitchen sink with a domestic dishwasher?
(A) 3 fixture units & 1/2" pipe
(B) 2.5 fixture units & 1/2" pipe
(C) 1.5 fixture units & 1/2" pipe
(D) None of these

69. A type of plastic water piping that may be tested with air is _____.
(A) PVC
(B) CPVC
(C) PE
(D) PEX

70. An expansion tank is not required for an instantaneous non-storage water heater.
(A) True
(B) False

CHAPTER 7
SANITARY DRAINAGE

The purpose of this chapter is to familiarize you with the requirements of conveying the discharge from plumbing fixtures to the public sewer or private sewage disposal system.

Some of the requirements found in Chapter 7 pertain to the type of material used, the types of fittings permitted to change directions, cleanout location, minimum slope for horizontal piping, and the correct sizing for the drainage and venting systems, including the building sewer.

1. The maximum fixture unit loading allowed on a 2 inch waste stack is _____.
   (A) four
   (B) six
   (C) eight
   (D) sixteen

2. Three-inch horizontal drainage piping shall have a grade or slope of not less than _____.
   (A) 1/4 inch per foot
   (B) 1/8 inch per foot
   (C) 1/2 inch per foot
   (D) 1 inch per foot

3. The maximum trap loading for a 3 inch trap is _____.
   (A) one unit
   (B) three units
   (C) four units
   (D) six units

4. The minimum sizes of vertical and horizontal drainage piping shall be determined from the _____.
   (A) location of fixtures
   (B) type of building
   (C) total bathrooms
   (D) total fixture units

5. The minimum size trap and trap arm for a bathtub is _____.
   (A) 1-1/2 inches
   (B) 2 inches
   (C) 2-1/2 inches
   (D) 3 inches

6. The maximum discharge capacity for intermittent flow only, in gallons per minute, for one fixture unit is equal to _____.
   (A) 7-1/2 gpm
   (B) 15 gpm
   (C) 30 gpm
   (D) 50 gpm

7. The minimum size drain line for a wall-hung urinal is _____ inches.
   (A) 2
   (B) 1-1/4
   (C) 1-1/2
   (D) 3

8. The acceptable fitting for connecting a horizontal drain line with another horizontal drain line is a _____.
   (A) sanitary tee
   (B) cross tee
   (C) combination wye and 1/8 bend
   (D) side inlet 1/4 bend

9. The clearance required in front of a 2 inch cleanout shall not be less than _____.
   (A) 6 inches
   (B) 12 inches
   (C) 18 inches
   (D) 24 inches

10. The discharge line from an ejector shall be provided with a backwater valve, ______.
    (A) or swing check valve and gate valve
    (B) or double check valve and globe valve
    (C) spring-loaded check valve, and union
    (D) gate valve, and flange

11. The minimum size discharge pipe from an ejector or pump in a single dwelling unit having a water closet connected thereto is ______.
    (A) 1-1/2 inches
    (B) 2 inches
    (C) 3 inches
    (D) 4 inches

12. A horizontal drain pipe has an existing load of 30 fixture units. After adding a new discharge pipe from an ejector rated at 60 gpm, the total load on the horizontal drain pipe will be ______.
    (A) 90 fixture units
    (B) 120 fixture units
    (C) 130 fixture units
    (D) 150 fixture units

13. Because of structural conditions, a 4-inch building drain is installed with a slope of 1/8 inch per foot. This installation will result in ______.
    (A) separation of solids and liquid
    (B) increased scouring and cleansing action
    (C) a reduction in fixture unit loading allowance
    (D) an increased allowance in fixture unit loading

For questions 14 through 21, refer to sketch below.

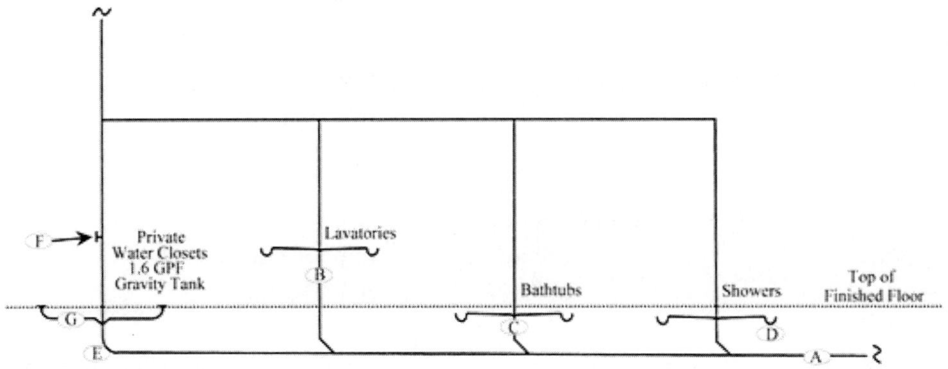

14. The total number of fixture units at "A" is ______.
    (A) 16
    (B) 20
    (C) 22
    (D) 24

15. The minimum size pipe at "A" is ______.
    (A) 2 inches
    (B) 3 inches
    (C) 4 inches
    (D) 5 inches

16. The minimum size pipe at "B" is ______.
    (A) 1-1/2 inches
    (B) 2 inches
    (C) 1-1/4 inches
    (D) 2-1/2 inches

17. The number of fixture units at "C" is ______.
    (A) 2
    (B) 4
    (C) 6
    (D) 8

18. The minimum size trap at "D" is _____.
    (A) 1-1/2 inches
    (B) 2 inches
    (C) 3 inches
    (D) 4 inches

19. The appropriate DWV PVC fitting at "E" would be a _____.
    (A) drainage wye
    (B) drainage tee
    (C) combination Y and 1/16 bend
    (D) long sweep 90° bend

20. The cleanout opening at "F" shall be at least _____.
    (A) 2 inch
    (B) 2-1/2 inch
    (C) 3 inch
    (D) 4 inch

21. The minimum size closet arm at "G" is _____ inch.
    (A) 2
    (B) 2-1/2
    (C) 3
    (D) 4

22. Except in single-family residences, drainage stacks 3 stories or more in height serving _____ are considered suds-producing fixtures.
    (A) kitchen sinks
    (B) bathtubs
    (C) washing machine standpipes
    (D) all of the above

23. Drainage connections shall not be made into a drainage piping system within _____ of any vertical to horizontal change of direction of a stack containing suds-producing fixtures.
    (A) 2 feet
    (B) 4 feet
    (C) 6 feet
    (D) 8 feet

24. Vertical drainage lines connecting with horizontal drainage lines shall enter through a _____.
    (A) 45 degree wye
    (B) 90 degree tee
    (C) combination wye and 1/8 bend
    (D) A and C apply

25. The maximum fixture units allowed on a 4-inch horizontal drain with a 1/8 inch per foot slope is _____.
    (A) 250 fixture units
    (B) 225 fixture units
    (C) 215 fixture units
    (D) 173 fixture units

26. The minimum size of any building sewer shall be determined on the basis of the _____.
    (A) type of building
    (B) total number of fixture units and building drain size
    (C) the material used for sewer piping
    (D) number of bedrooms in the building

27. A straight run of building sewer will require cleanouts at intervals not to exceed _____.
    (A) 50 feet
    (B) 75 feet
    (C) 100 feet
    (D) 300 feet

28. With appropriate cause and Authority Having Jurisdiction approval, building sewers 8 inches or larger shall be permitted to have a slope of not less than _____.
    (A) 1/16 inch per foot
    (B) 1/8 inch per foot
    (C) 1/4 inch per foot
    (D) 1/2 inch per foot

29. The maximum distance between manholes shall not exceed ______.
(A) 50 feet
(B) 100 feet
(C) 200 feet
(D) 300 feet

30. ABS and PVC DWV piping may be installed in structures of ______.
(A) noncombustible construction
(B) unlimited height
(C) combustible construction
(D) all of these

31. Fittings on screwed pipe shall be of (the) ______.
(A) caulk-drainage type
(B) duro iron
(C) recessed-drainage type
(D) galvanized steel

32. Except for plastic pipe, in lieu of a water test, drainage piping may be tested with ______.
(A) air
(B) ether
(C) smoke
(D) gas

33. Threads in drainage fittings shall be tapped to allow for ______.
(A) adjustment
(B) grade
(C) expansion
(D) economy

34. The maximum number of fixture units allowed on a 4-inch building sewer with a slope of 1/8 inch per foot, as compared to a 4-inch building sewer with a slope of 1/4 inch per foot is ______.
(A) the same
(B) less
(C) more
(D) unlimited

35. Macerating toilet systems shall be permitted ______.
(A) for private use
(B) for public use
(C) when approved by Authority Having Jurisdiction
(D) not allowed by code

36. Sumps for macerating toilets shall be ______.
(A) above ground
(B) a minimum of 10 gallons
(C) watertight and gastight
(D) in the same room as the toilet

37. The drainage fixture unit value assigned to a bathtub or combination bath/shower is ______.
(A) 1.5
(B) 2.0
(C) 2.5
(D) 3.0

38. With a few exceptions, all horizontal drainage piping shall be provided with a cleanout at ______.
(A) each 45 degree fitting
(B) its upper terminal
(C) 25 foot intervals
(D) each offset

39. Every abandoned building sewer shall be plugged or capped ______.
(A) at the sidewalk or curb
(B) within 5 feet of the property line
(C) at the street tap
(D) where convenient

40. The building sewer and water piping may be installed in the same trench (with the proper clearances) if the building sewer piping is _____.
(A) cast iron
(B) extra strength vitrified clay
(C) PVC
(D) any of the above

41. Cleanouts shall be designed to be _____.
(A) threaded
(B) gastight and watertight
(C) easy opening
(D) concealed

42. A domestic dishwasher, with independent drain, has a drainage fixture unit value of _____.
(A) 2
(B) 3
(C) 4
(D) 1

43. After caulking a lead joint in a cast iron hub, the finished joint shall not exceed _____ of an inch below the rim of the hub.
(A) 1/8
(B) 1/6
(C) 1/5
(D) 1/4

44. A connection to the public sewer, or private sewage disposal system, is required for _____.
(A) buildings with or without plumbing fixtures
(B) buildings with plumbing fixtures
(C) buildings with emergency drain systems
(D) buildings with storm drain systems

45. An acceptable method for joining Polyethylene sewer pipe is _____.
(A) threaded
(B) butt fusion
(C) solvent cement
(D) push on

46. Replacement of existing building sewer and building storm sewers using trenchless methodology and materials shall be installed in accordance with ASTM F1296, ASTM F2561, ASTM F2599, or ASTM F3240.
(A) True
(B) False

47. Dual sewage ejectors for public use require _____ alarm.
(A) audio
(B) audio and visual
(C) visual
(D) audio or visual

CHAPTER 8
INDIRECT WASTES

In restaurants, bars, food establishments, and many other commercial-type buildings where it is necessary to protect against the possibility of backflow and contamination, an indirect waste is a method of discharging the waste pipe through an airgap or airbreak into an open receptor. This eliminates a possible cause of food contamination from the sanitary drainage system. For these, the Uniform Plumbing Code permits a special system of waste piping, which is regulated by this chapter.

Drains from cooling counters, cold storage rooms, refrigerators, etc., including drains from fixtures used for handling food that may be consumed with further cooking, are required to be drained by means of indirect waste piping terminating into an approved receptor.

This examination will test you on your knowledge of the above regulations, including regulations for clothes washer standpipes, appliances, steam and hot water condenser and intercepting tank, along with chemical and industrial wastes.

1. All indirect waste piping that requires an airgap shall discharge into the building drainage system through ______.
   (A) an automatic vent
   (B) an airgap of two pipe diameters
   (C) a drainage airgap of 1 inch minimum
   (D) a trap

2. Except for refrigeration coils and ice-making machines, the minimum size of the indirect waste pipe shall not be smaller than the drain on the unit and not smaller than ______ inch(es).
   (A) 3/4
   (B) 1
   (C) 1-1/4
   (D) 1-1/2

3. Floor drains in walk-in coolers shall discharge ______.
   (A) directly to the main building drain
   (B) directly to the exterior of the building
   (C) to a separate drainage line discharging into an outside receptor through an airgap or airbreak
   (D) to a grease interceptor

4. Drains, overflows, or relief pipes from a water distribution system shall discharge to the building drain by ______.
   (A) indirect waste by means of a water-distribution airgap
   (B) direct connection to the building drain
   (C) direct connection to the building drain through a trap
   (D) indirect waste piping through a vented trap

5. Stills, sterilizers, and similar equipment requiring waste connections and used for sterile materials shall be ______.
   (A) connected directly to the building drain
   (B) directly connected to the building drain using an airbreak
   (C) indirectly connected by means of an airgap
   (D) indirectly connected to a lavatory tailpiece

6. Devices that are not regularly classified as plumbing fixtures, but which have drip or drainage outlets, shall be drained by _____.
   (A) a connection directly to the building
   (B) indirect waste pipes discharging into an open receptor through an airgap or airbreak
   (C) indirect waste piping connected to the building drain
   (D) direct connection to a storm sewer

7. No plumbing fixture that is receiving discharge from indirect waste pipes shall be installed until first approved by the _____.
   (A) architect
   (B) local health department
   (C) Authority Having Jurisdiction
   (D) mechanical engineer

8. The developed length from the fixture outlet of a bar sink to the receptor shall not exceed _____ feet.
   (A) 6
   (B) 5
   (C) 3
   (D) 15

9. All receptors receiving the discharge from indirect waste piping shall be _____.
   (A) approved for the proposed use
   (B) shaped and sized to prevent splashing or flooding
   (C) located where they are readily accessible for inspection and cleaning
   (D) all of the above

10. A standpipe for a clothes washer shall have a minimum and maximum length of _____.
   (A) 4 inches and 12 inches
   (B) 18 inches and 30 inches
   (C) 18 inches and 24 inches
   (D) 12 inches and 36 inches

11. An indirect waste receptor shall not be installed in a _____.
   (A) toilet room
   (B) boiler room
   (C) commercial kitchen
   (D) air compressor equipment room

12. No drain, overflow, or relief vent from a water supply system, or any other discharges under pressure shall be discharge to the drainage system _____.
   (A) by means of indirect waste through an airgap or airbreak
   (B) by direct connection
   (C) through the garbage disposal
   (D) with a branch wye tailpiece

13. A domestic dishwashing machine shall be connected to the drainage system by using _____.
   (A) a direct connection
   (B) a double-check valve
   (C) an approved airgap fitting
   (D) a Hartford Loop

14. Drinking fountains may be installed with _____.
   (A) a hose connection
   (B) a condensate pump
   (C) a Hartford Loop
   (D) indirect waste

15. Water having a temperature above _____ shall not be discharged under pressure directly into the drainage system.
   (A) 212°F
   (B) 180°F
   (C) 140°F
   (D) 232°F

16. Every indirect waste interceptor receiving discharge containing particles that would clog the receptor drain shall _____.
    (A) have a readily removable beehive strainer
    (B) drain into a funnel connected to the floor sink
    (C) be connected directly into the drainage system
    (D) both A and B

17. _____ shall not be used for chemical or industrial waste that requires pretreatment.
    (A) Glass
    (B) PVC
    (C) Cast-iron
    (D) Copper tube

18. Lead waste pipe that may receive discharge containing acid or corrosive chemicals, and each vent pipe connected to it, shall be not less than _____ wall thickness.
    (A) 1/16 inch
    (B) 3/32 inch
    (C) 1/8 inch
    (D) 5/32 inch

19. Whenever practicable, all piping receiving chemical waste shall be _____.
    (A) concealed
    (B) readily accessible
    (C) covered
    (D) in clear view

20. The owner shall make and keep a _____.
    (A) blueprint of the construction
    (B) log of all chemicals used
    (C) permanent record of the location of piping and venting carrying chemical waste
    (D) record of materials used for all chemical wastes

21. No chemical waste vent shall _____.
    (A) terminate outside the building
    (B) be connected to any vent from water service lines
    (C) intersect vents for other services
    (D) connect to any waste line

22. No chemical waste shall be discharged without _____.
    (A) being properly vented
    (B) approval of the local Authority Having Jurisdiction
    (C) first going through a chemical waste separator
    (D) being properly trapped

23. Water lifts, expansion tanks, cooling jackets, sprinkler systems, drip or overflow pans, and other similar devices that discharge clear wastewater into the building drainage system shall _____.
    (A) be directly trapped
    (B) be discharged into an interceptor
    (C) discharge through an indirect waste
    (D) be individually vented

24. Pipes carrying wastewater from swimming or wading pools, including pool drainage, backwash from filters, and discharge from pumps, shall be _____.
(A) directly connected to the building drain and properly trapped and vented
(B) connected to an interceptor
(C) connected to a holding tank
(D) installed as an indirect waste

25. When discharged into the drainage system, condensate shall drain by means of a (an) _____.
(A) indirect waste pipe
(B) direct waste pipe
(C) Johnson Tee
(D) backflow preventer

26. Condensate or wastewater shall not drain over _____.
(A) dry wells
(B) a public way
(C) leach pits
(D) the tailpiece of plumbing fixtures

27. Condensate from fuel-burning condensing appliances shall be collected and discharged _____.
(A) to a planter box
(B) to a public way
(C) to an approved plumbing fixture or disposal area
(D) directly connect to the building sewer

28. If the trap from a sink located in a bar cannot be vented, _____.
(A) the drain shall discharge through an airgap or airbreak into an approved receptor that is vented
(B) the fixture must be relocated
(C) the discharge may be pumped to an approved receptor
(D) a drum trap shall be used

29. The minimum size condensate waste pipe from a 12-ton refrigeration unit is _____.
(A) 3/4 inch
(B) 1 inch
(C) 1-1/4 inches
(D) 1-1/2 inches

30. Air conditioning condensate waste shall connect indirectly through an air gap or air break to a _____.
(A) floor drain
(B) plumbing fixture tailpiece
(C) leach pit
(D) all the above

CHAPTER 9
VENTS

The Uniform Plumbing Code requires that, when practical, all plumbing fixtures shall drain to the public sewer or private sewer by gravity. To accomplish this, certain conditions have to be taken into consideration. The same pressure must exist within the drainage system as the air pressure outside of such system. This means that openings have to be provided in specific locations throughout the drainage system. These openings are intended to provide a means for preventing variations in air pressures, plus or minus, within the drainage system. This variation is caused by the flow of liquid wastes and sewage.

Fixture traps are required to have a liquid seal that is very sensitive to pressure variations in and out of a drainage system. For this reason, the code requires the installation of vent pipes. They protect the traps and their water seals against the effects of "back pressure" and "siphonage," which are, respectively, positive and negative pressures.

The principles of horizontal and vertical wet venting on one common waste and vent pipe is practical for fixtures set within the distance of their trap arm limitations from the common waste and vent.

Combination waste and vent systems are permitted where structural conditions preclude the installation of conventional systems. This type of system is essential in buildings where individual venting of floor sinks, floor drains, or showers is impractical.

The questions in this examination will test you on your knowledge of code requirements for venting systems. Study Table 703.2 and Chapter 9 in the code prior to taking this examination. Review any questions you answered wrong before moving on to the next chapter.

1. Vent pipes protect fixture traps against ______.
   (A) evaporation
   (B) siphonage
   (C) clogging
   (D) cross-connections

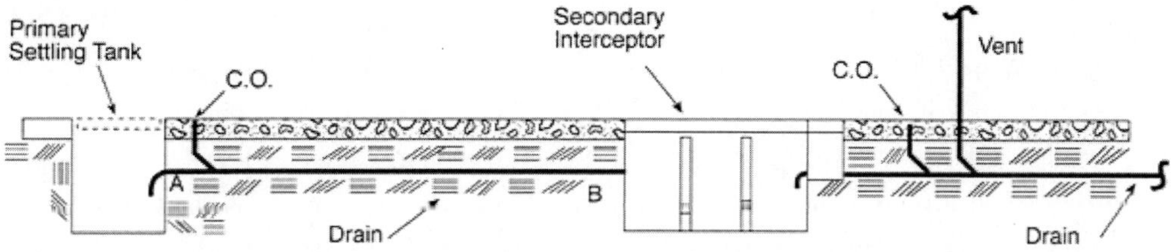

2. In the above sketch, the primary settling tank ______.
   (A) requires a backwater valve
   (B) requires a vent pipe
   (C) is correct as shown
   (D) does not require a cleanout

3. Under certain conditions, traps serving sinks that are part of the equipment of bars, soda fountains, and counters ______.
   (A) need not be vented
   (B) shall be vented
   (C) may be wet-vented
   (D) need relief vents

4. One factor for determining the size of a vent pipe is ______.
   (A) its accessibility
   (B) its length
   (C) the type of building
   (D) the price of material

5. The combination of vent pipes that will provide sufficient aggregate cross-sectional vent area for a 4-inch house sewer is ______ vent pipes.
   (A) 1 – 2 inch and 4 – 1-1/2 inch
   (B) 2 – 2 inch and 1 – 2-1/2 inch
   (C) 3 – 2 inch and 1 – 1-1/2 inch
   (D) 4 – 2 inch

6. The maximum horizontal length permitted for a required 2-inch vent pipe is ______.
   (A) 20 feet
   (B) 40 feet
   (C) 60 feet
   (D) 120 feet

7. The diameter of an individual vent shall not be less than ______.
   (A) 1-1/4 inches
   (B) 1-1/2 inches
   (C) 2 inches
   (D) 3 inches

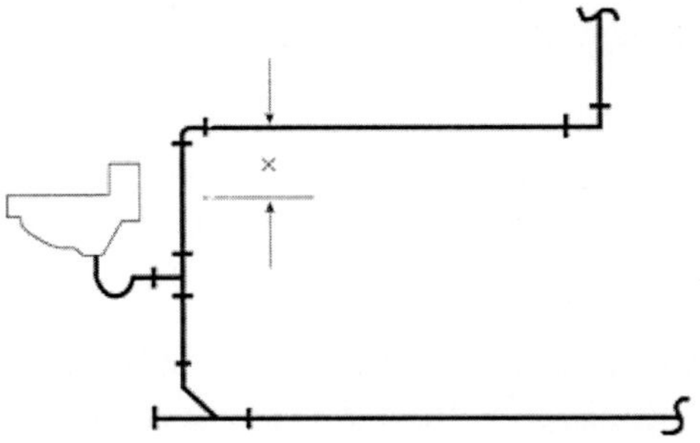

8. The distance "X" in the sketch above shall be at least ______.
   (A) 1 inch
   (B) 2 inches
   (C) 6 inches
   (D) 12 inches

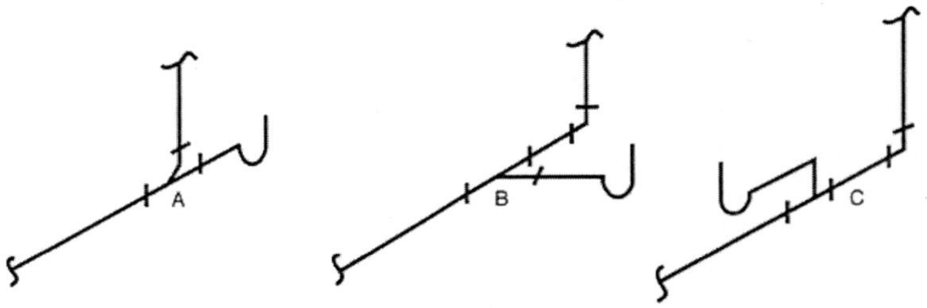

9. The acceptable method for connecting a vent pipe to a horizontal drain pipe is shown above by ______.
   (A) sketch "A"
   (B) sketches "A" and "B"
   (C) sketch "B"
   (D) sketches "B" and "C"

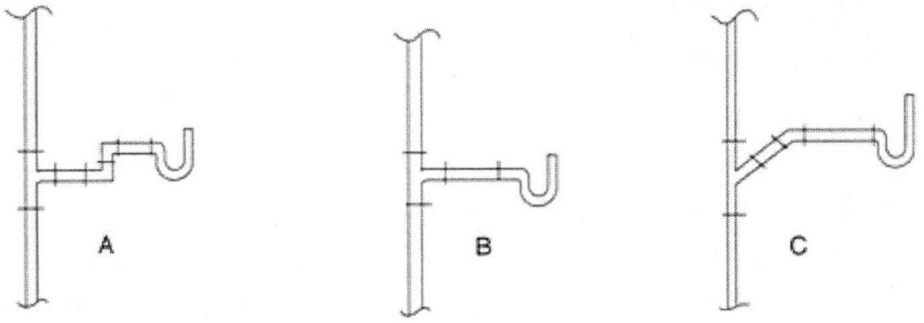

10. In the above sketch, the vent pipe opening is above the crown weir of the trap in ______.
    (A) sketch "A"
    (B) sketch "B"
    (C) sketch "C"
    (D) sketches "A" and "B"

11. A 2-inch vent pipe will serve a total of ______.
    (A) 14 bathtubs
    (B) 16 showers
    (C) 24 lavatories
    (D) 15 water closets

12. Each vent pipe or stack shall extend through its flashing and shall terminate vertically above the roof at least ______.
    (A) 6 inches
    (B) 12 inches
    (C) 18 inches
    (D) 24 inches

13. Refer to question 12. The distance from the vent pipe opening to any vertical surface shall be at least _____.
    (A) 6 inches
    (B) 12 inches
    (C) 18 inches
    (D) 24 inches

14. Horizontal termination of vents from any openable window, door opening, air intake or vent shaft shall be not less than _____.
    (A) 1 foot
    (B) 2 feet
    (C) 10 feet
    (D) 12 feet

15. Except for alleys and streets, vent termination from a lot line shall not be less than _____.
    (A) 1 foot
    (B) 2 feet
    (C) 3 feet
    (D) 4 feet

16. Vents terminating in roof areas used for sunbathing shall be extended above the roof at least _____.
    (A) 7 feet
    (B) 10 feet
    (C) 12 feet
    (D) 15 feet

17. Joints at the roof around vent pipes shall be made watertight by the use of _____.
    (A) waterproof epoxy sealer
    (B) three layers 15 pound felt
    (C) hot tar compound
    (D) approved flashings

18. Parallel vent stacks are required for drainage stacks that extend _____.
    (A) five stories or more
    (B) 10 stories or more
    (C) 50 feet
    (D) 100 feet

19. The parallel vent stack shall be connected to the drainage stack by means of a _____.
    (A) circuit vent
    (B) loop vent
    (C) yoke vent
    (D) continuous vent

20. The pipe size for the continuous vent serving lavatories installed back-to-back shall be not less than _____ inch(es).
    (A) 1
    (B) 1-1/4
    (C) 1-1/2
    (D) 2

21. Horizontal vent pipes shall be level or so graded as to drip back by gravity to _____.
    (A) the main vent pipe
    (B) the main soil stack
    (C) the loop vent
    (D) the drainage pipe they serve

22. The minimum size vent pipe that may serve a 4-inch trap is _____.
    (A) 1-1/4 inches
    (B) 1-1/2 inches
    (C) 2 inches
    (D) 4 inches

23. Increasing vents one pipe size will _____.
    (A) provide increased drainage
    (B) allow a uniform installation
    (C) remove lengths limitations
    (D) allow vents to be shorter

24. Changes in direction of vent piping shall be made by _____.
    (A) heating the pipe and bending
    (B) welded joints where possible
    (C) threaded fittings or caulked joints
    (D) appropriate use of approved fittings

25. A drainage stack extends 14 stories above the building drain and is served by a parallel vent stack. A relief yoke vent connection between the vent stack and the drainage stack is required at ______.
    (A) floors 5 and 10
    (B) floors 4 and 9
    (C) floors 4 and 10
    (D) floors 5 and 9

26. The fitting that is required for making the yoke vent intersection with the drainage stack is a ______.
    (A) sanitary tee
    (B) tapped tee
    (C) double tee branch
    (D) wye branch

27. Referring to question 26 above, the size of the yoke vent shall be ______.
    (A) not less in diameter than either the drainage or the vent stack, whichever is smaller
    (B) a minimum of 4 inches
    (C) two pipe sizes larger than the drain line
    (D) two pipe sizes smaller than the drain line

28. If, due to structural conditions, horizontal vents are less than 6 inches above the overflow rim of the fixture, the horizontal portion shall be installed with ______.
    (A) approved venting material and fittings
    (B) approved drainage material and fittings and grade to drain
    (C) smaller diameter piping
    (D) larger diameter piping

29. Vent terminals in locations having minimum design temperatures below 0°F shall be a minimum of ______ inches.
    (A) 1-1/2
    (B) 2
    (C) 3
    (D) 4

30. Copper tube shall not be used for ______.
    (A) chemical or industrial wastes
    (B) water piping
    (C) underground drainage piping
    (D) aboveground drainage piping

31. ABS or PVC DWV piping installations exposed within ducts or plenums shall be limited to a flame-spread index of not more than 25 or a smoke developed index of 50 except for ______.
    (A) two-story buildings
    (B) structures not over three floors above grade
    (C) the restrictions found in IS 31
    (D) single-family dwelling units

32. Sheet lead used for flashings of vent terminals shall be not less than ______.
    (A) 2 pounds per square foot
    (B) 3 pounds per square foot
    (C) 4 pounds per square foot
    (D) 5 pounds per square foot

Refer to sketch below for questions 33 through 37.

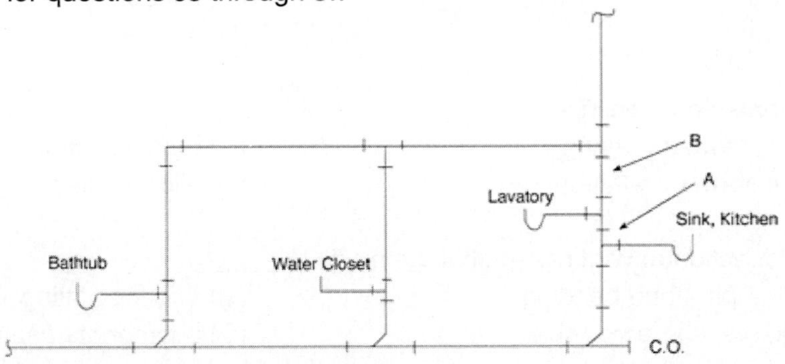

33. Section "A" refers to _____.
(A) a combination waste and vent system
(B) a vertical wet-vented section
(C) special venting for island fixtures
(D) an indirect waste and vent system

34. The pipe section between the kitchen sink inlet and the lavatory inlet (point "A") serves as a _____.
(A) vent pipe for the kitchen sink
(B) waste pipe for the lavatory
(C) vent pipe for the lavatory
(D) both A and B

35. The total number of fixture units at point "B" is _____.
(A) one unit
(B) two units
(C) three units
(D) four units

36. The vent section "B" shall be a minimum pipe size of _____.
(A) 1-1/4 inch
(B) 1-1/2 inch
(C) 2 inch
(D) 2-1/2 inch

37. In this type of installation, the diameter of the lavatory trap arm _____.
(A) shall be increased one pipe size
(B) shall be increased two pipe sizes
(C) may be as normally required
(D) may be decreased one pipe size

Refer to the sketch below for questions 38 through 43.

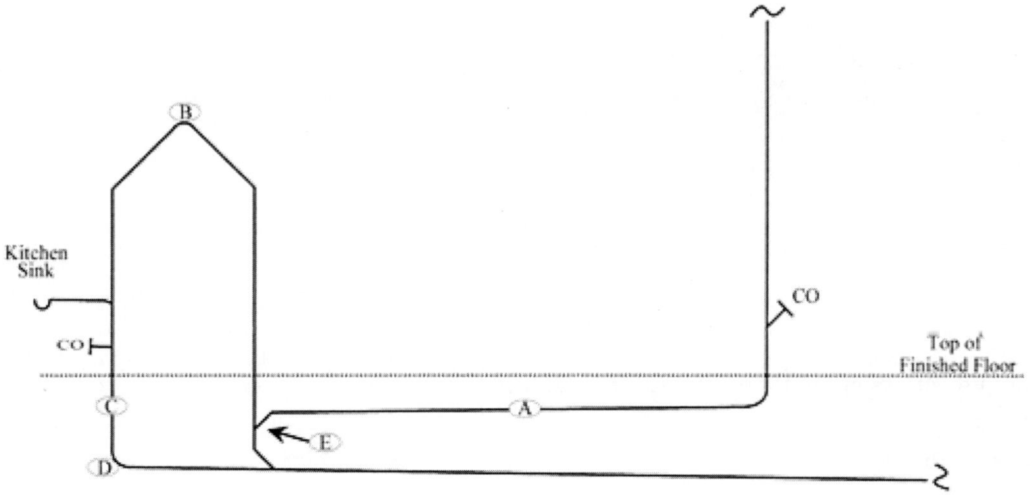

38. The sketch represents _____.
(A) reverse circuit loop venting
(B) continuous waste and vent
(C) special venting for island fixtures
(D) combination waste and vent

39. The pipe section at "A" is called a _____.
(A) foot vent
(B) yoke vent
(C) loop vent
(D) fixture vent

40. The fitting at "B" is a _____.
(A) 45-degree elbow
(B) 90-degree elbow
(C) 1/16 bend
(D) 1/8 bend

41. The minimum size (diameter) at “C” is ______.
    (A) 1-1/2 inches
    (B) 2 inches
    (C) 2-1/2 inches
    (D) 3 inches

42. The type of fitting acceptable at “D” is a ______.
    (A) sanitary tee
    (B) sanitary tapped tee
    (C) 90-degree elbow
    (D) 90-degree sweep

43. Of the following fittings, the one that would be approved at “E” is the ______.
    (A) 1-1/2 inch wye branch fitting
    (B) 1-1/2 inch sanitary tee
    (C) 2 inch tapped tee
    (D) 2 inch sink tee

Refer to the sketch below for questions 44 through 49.
(Additional information may be found in Appendix B of the UPC)

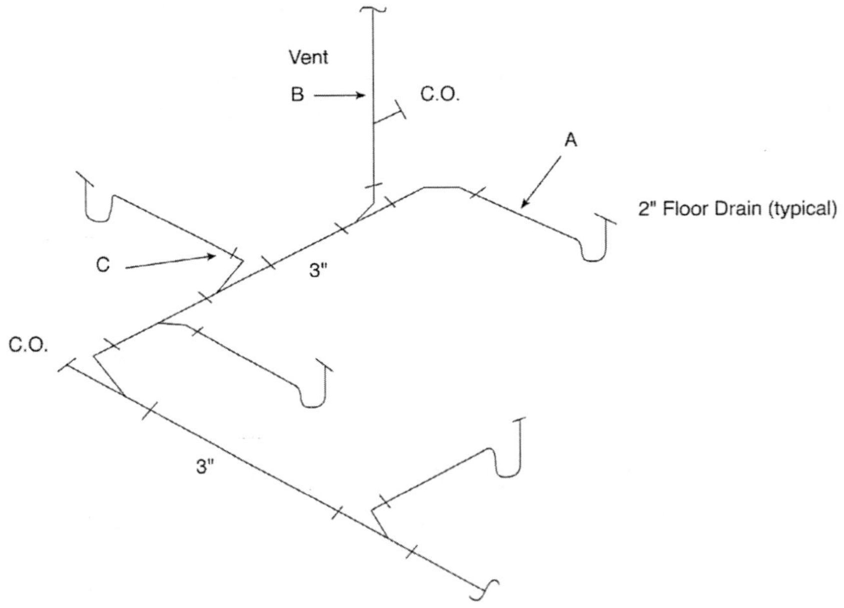

44. The maximum length for piping “C” without a vent is ______.
    (A) 5 feet
    (B) 10 feet
    (C) 15 feet
    (D) 25 feet

45. The minimum size (diameter) of the tailpiece for the floor drain served by “A” is ______.
    (A) 1-1/2 inches
    (B) 2 inches
    (C) 2-1/2 inches
    (D) 3 inches

46. The minimum size (diameter) for branch “A” is ______.
    (A) 2 inches
    (B) 2-1/2 inches
    (C) 3 inches
    (D) 4 inches

47. The minimum size (diameter) for vent piping “B” shall be approximately equal to ______.
    (A) the cross-sectional area of the drain pipe served
    (B) one-half the inside cross-sectional area of the drain pipe served
    (C) the size of the uppermost trap
    (D) one-half the size of the uppermost trap

48. Long mains shall be provided with relief vents at intervals of _____.
    (A) 50 feet
    (B) 75 feet
    (C) 100 feet
    (D) 300 feet

49. The vent connection "B" shall be _____.
    (A) downstream of the last fixture
    (B) within 15 feet of the fixture discharging the most total volume of waste
    (C) within 15 feet of the uppermost fixture
    (D) taken off above the centerline of the drain pipe and downstream of the trap being served

50. Horizontal wet venting _____.
    (A) is not permitted by the code
    (B) shall be permitted for public bathrooms
    (C) shall be permitted for a bathroom group located on the same floor level
    (D) shall be a minimum 4 inches in size

51. Of the options shown, the fixture that serves as a dry vent connection to the horizontal wet vent is a _____.
    (A) kitchen sink
    (B) laundry tub
    (C) shower
    (D) bar sink

52. The vent system shall be designed to prevent a trap seal from being exposed to a pressure differential on the outlet side of the trap that exceeds _____ inch of a water column.
    (A) 0.50
    (B) 1
    (C) 2
    (D) 3

53. A horizontal wet vent shall be not less than two inches in diameter for _____ dfu or less.
    (A) 2
    (B) 4
    (C) 5
    (D) 8

54. The dry vent connection to the wet vent shall be a (an) _____.
    (A) relief vent
    (B) yoke vent
    (C) foot vent
    (D) individual vent

55. The _____ fixture drain or trap arm connection to the wet vent shall be downstream of the fixture drain or trap arm connections to the horizontal branch.
    (A) shower
    (B) floor drain
    (C) water closet
    (D) bathtub

56. An engineered vent system shall be designed by _____ and approved in accordance with Section 301.5.
    (A) a registered design professional
    (B) the building owner
    (C) the plumber
    (D) an engineer

57. What is the maximum number of fixtures permitted on a circuit vent to be connected to a horizontal branch drain?
    (A) 4
    (B) 6
    (C) 8
    (D) 10

58. In circuit venting the horizontal branch shall be classified as a _____ from the most downstream trap arm connection to the most upstream trap arm connection to the horizontal branch.
    (A) cleanout
    (B) drain
    (C) vent
    (D) drain and vent

59. When circuit venting, fixtures discharging to a relief vent shall be one or two fixtures unit fixtures but shall not exceed a total of _____ fixture units.
(A) 3
(B) 5
(C) 4
(D) 6

60. What is the minimum size circuit vent?
(A) 1-1/4"
(B) 1-1/2"
(C) 2"
(D) 3"

61. When circuit venting, the vent section of the horizontal branch drain shall slope not more than _____.
(A) 1/8" per foot
(B) 1/4" per foot
(C) 1/2" per foot
(D) 1" per foot

62. Parallel horizontal circuit vented branches shall be permitted to connect on _____.
(A) the same floor level
(B) only the first and fifth floor
(C) only between first and second floor
(D) only through increasing the pipe size four times

63. A 2 inch relief vent shall be provided for circuit-vented horizontal branches receiving the discharge of _____ or more water closets when connecting to a drainage stack receiving soil and waste discharge from upper branches.
(A) three
(B) four
(C) six
(D) eight

64. Fixtures, other than circuit vented fixtures, located on the same floor as circuit-vented fixtures shall be either common vented or _____.
(A) individually vented
(B) wet vented
(C) island vented
(D) stack vented

65. Where exposed to sunlight, ABS and PVC vent piping penetrating the roof shall be protected by _____.
(A) oil based paint
(B) latex based paint
(C) acrylic based paint
(D) roofing tar

66. Under what conditions are multiple buildings allowed to 'share' aggregate cross-sectional venting areas to satisfy the requirements of the total vent pipe areas being not less than the area of the largest required building sewer?
(A) one ownership
(B) common building sewer
(C) buildings occupy the same lot
(D) A, B & C are all required

67. A bathroom group is made up of:
(A) water closet, tub/shower and lavatory
(B) bidet, lavatory and shower
(C) shower, bidet and emergency floor drain
(D) all of the above meet the definition

68. When is the sidewall venting of DWV permitted?
(A) weather conditions are extreme
(B) sidewall venting is not permitted
(C) roofing is cement tile
(D) roof pitch is not less than 8/12

69. Branches serving traps in a combination waste and vent system shall connect to the main line at an angle not exceeding _____ percent.
(A) 2.0
(B) 5.0
(C) 22.5
(D) 50.0

CHAPTER 10
TRAPS AND INTERCEPTORS

A review of definitions for traps and interceptors in the Uniform Plumbing Code reveals that they perform an entirely different function with respect to each other.

Traps are designed to provide a liquid seal that will prevent back passage of sewer gas. However, in operation, they are to be self-cleaning and, therefore, not specifically designed to trap anything but the liquid water seal. In contrast to the foregoing, interceptors are not self-cleaning and, in fact, are designed to separate and retain deleterious, hazardous, or undesirable matter from normal wastes.

Auto-wash racks require the installation of interceptors to retain sand, grease, and other matter. Restaurants and similar establishments that discharge an excessive amount of grease into the drainage system are required to install grease interceptors. The regulations for these and other types of industrial separators is the subject of this examination.

Traps that were used in plumbing installations at one time and found unsuitable are prohibited by the code. These are listed in this part of the code. The proper depth of the water seal in traps should be known, including provisions for maintaining this water seal in traps that are subject to infrequent use.

1. Each plumbing fixture, except for those having integral traps, shall be _____.
   (A) separately trapped by an approved liquid seal trap
   (B) drained by means of an indirect waste line
   (C) connected directly to the building drain without a trap
   (D) illegal and should not be used

2. One trap may serve a set of not more than _____ single compartment sinks.
   (A) one
   (B) two
   (C) three
   (D) four

3. A food waste disposal unit installed in restaurant, commercial, or industrial sinks _____.
   (A) may share a common trap
   (B) shall be separately trapped
   (C) should be indirectly wasted
   (D) need not be trapped

4. The maximum vertical distance for a tailpiece allowed between a fixture outlet and the trap weir is _____.
   (A) 6 inches
   (B) 12 inches
   (C) 18 inches
   (D) 24 inches

5. Each plumbing fixture trap shall be protected against siphonage and backpressure by a _____.
   (A) sanitary tee
   (B) vent pipe
   (C) air admittance device
   (D) backflow preventer

6. A trap arm smaller than 3 inches may change direction without a cleanout if the change in direction does not exceed _____.
   (A) 30 degrees
   (B) 45 degrees
   (C) 90 degrees
   (D) 180 degrees

7. The vent pipe opening from a soil or waste pipe, except for water closets and similar fixtures, _____.
   (A) may be below the trap weir
   (B) shall not be below the weir of the trap
   (C) must be below the weir of the trap
   (D) does not need a vent

8. The developed length of a 2" trap arm, from the trap weir to the inner edge of the vent shall not exceed _____ feet.
   (A) 3
   (B) 5
   (C) 6
   (D) 10

9. Each trap, except for traps within interceptors or similar devices, _____.
   (A) shall be self-cleaning
   (B) shall be one pipe size smaller than the fixture outlet
   (C) shall have a cleanout on the bottom of the trap
   (D) need not be vented

10. No more than _____ approved slip joint fitting(s) may be used on the outlet side of a trap.
   (A) one
   (B) two
   (C) three
   (D) four

11. The size of the trap connected to a fixture shall be sufficient to drain the fixture rapidly but _____.
   (A) in no case smaller than the fixture drain
   (B) not more than one size larger than the fixture drain
   (C) shall be the same size as the trap arm to which it is connected
   (D) all of the above

12. No trap shall be used that _____.
   (A) depends on movable parts for its seal
   (B) has, with some exceptions, concealed interior partitions
   (C) is classified as an S-trap or bell trap
   (D) all of the above

13. Each fixture trap shall have a liquid seal of _____.
   (A) not less than 2 inches
   (B) not more than 4 inches
   (C) both A and B
   (D) none of the above

14. Floor drains shall connect into a trap so constructed that it can be _____.
   (A) accessible
   (B) easily removed
   (C) readily cleaned
   (D) none of the above

15. When deemed necessary, floor drains or similar traps directly connected to the drainage system and subject to infrequent use shall be _____.
   (A) primed weekly
   (B) covered when not in use
   (C) protected with an accessible trap seal primer
   (D) increased one trap size

16. Each building trap when installed shall be provided with _____.
   (A) a cleanout
   (B) a relieving vent or fresh air intake on the inlet side of the trap
   (C) both A and B
   (D) a trap primer

17. Interceptors shall be provided when, in the judgment of the Authority Having Jurisdiction, they are necessary for the proper handling of liquid wastes containing _____.
    (A) grease and flammable wastes
    (B) sand and/or solids
    (C) acid or alkaline substances
    (D) all of the above

18. Interceptors for sand and similar heavy solids shall have a water seal of not less than _____.
    (A) 4 inches
    (B) 6 inches
    (C) 8 inches
    (D) 10 inches

19. Interceptors shall _____.
    (A) be properly vented
    (B) have covers that are readily accessible for service and maintenance
    (C) be placed aboveground
    (D) both A and B apply

20. Every public or private wash rack and floor or slab used for cleaning machinery or machine parts shall _____.
    (A) be adequately protected against storm or surface water
    (B) drain or discharge into the main sewer system
    (C) drain or discharge into an approved interceptor
    (D) both A and C

21. Laundry equipment in commercial buildings that do not have integral strainers shall discharge into an interceptor having a wire basket or similar device that will prevent the passing of solids _____ or larger in maximum dimension.
    (A) 1/4 inch
    (B) 1/8 inch
    (C) 1/2 inch
    (D) 3/4 inch

22. The Authority Having Jurisdiction has mandated grease pretreatment for the following fixtures: four non-emergency floor drains, two commercial food prep sinks, one service sink with a 3-inch trap, and two dishwashers discharging 20 gallons per minute each. The minimum size gravity grease interceptor in gallons required shall be _____.
    (A) 500
    (B) 750
    (C) 1,000
    (D) 1,250

23. Each plumbing fixture connected to a type A and B hydromechanical grease interceptor shall be provided with a _____.
    (A) vent terminating to the atmosphere inside the building (turndown with screen)
    (B) flow control device
    (C) gate valve
    (D) removable cover

24. Each fixture discharging into a gravity grease interceptor shall be _____.
    (A) in the same room
    (B) individually trapped and vented
    (C) readily accessible
    (D) approved for that use

25. The total capacity in gallons of fixtures discharging into any hydromechanical grease interceptor shall not exceed _____ times the certified gpm flow rate.
(A) 1-1/2
(B) 2
(C) 2-1/2
(D) 3

26. Sand interceptors shall be required _____.
(A) whenever the discharge of a fixture or drain may contain solids or semi-solids heavier than water that would be harmful to the drainage system
(B) whenever the Authority Having Jurisdiction deems it advisable to protect the drainage system
(C) both A and B
(D) none of the above

27. The sand interceptor shall have a minimum dimension of _____ square feet for the net free opening of the inlet section and a minimum depth under the invert of the outlet pipe of _____ feet.
(A) 1-1/2, 1-1/2
(B) 2-1/2, 2-1/2
(C) 2, 2
(D) 3, 3

28. Sand and similar interceptors shall have a water seal of not less than _____ inches.
(A) 4
(B) 6
(C) 8
(D) 10

29. Repair garages and gas stations with grease racks or grease pits and all factories that have oily and/or flammable wastes shall be provided with an oil or flammable liquid interceptor that shall _____.
(A) be connected to necessary floor drains
(B) have a separation or vapor compartment independently vented to the outer air
(C) have a flammable vapor vent not less than 2 inches
(D) all of the above

30. Oil and flammable liquid interceptors not rated by the manufacturer shall have a depth of not less than _____ feet below the invert of the discharge drain and an outlet opening with not less than _____ inch water seal.
(A) 1-1/2, 12
(B) 2, 18
(C) 2-1/2, 24
(D) 3, 36

31. If the Authority Having Jurisdiction determines that a grease interceptor is not being properly cleaned, the Authority Having Jurisdiction has the authority to mandate _____.
(A) hiring more employees
(B) firing the manager
(C) a maintenance program
(D) less grease to be used

32. A _____ shall be installed downstream of hydromechanical grease interceptor in accordance with the requirements of this code.
(A) check valve
(B) vent
(C) backflow preventer
(D) filter

33. Gravity grease interceptors shall not be installed in a building where _____.
(A) laundry is handled
(B) food is handled
(C) there is public access
(D) the fixture served is on the second floor

34. The volume of the gravity interceptor shall be determined by using _____.
(A) Table 313.1
(B) Table 1014.2.1
(C) Table 1014.3.6
(D) Table 610.3

35. Gravity grease interceptors shall be placed as close as practical to the _____.
(A) fixtures they serve
(B) soil pipe
(C) system vents
(D) egress doors

36. It is recommended that a _____ be located at the outlet end of all gravity grease interceptors.
(A) pH monitor
(B) CO_2 monitor
(C) sight glass
(D) sample box

37. Abandoned grease interceptors shall be _____ and _____ as required for abandoned sewers and sewage disposal facilities.
(A) broke up, removed
(B) pumped, filled
(C) painted, sealed
(D) cleaned, reused

38. Water closets and urinals shall _____ into or through a grease interceptor.
(A) drain
(B) directly connect
(C) indirectly connect
(D) not drain

39. Listed _____ shall be permitted to be used to connect listed metal tubing traps.
(A) male adapters
(B) female adapters
(C) PEX adapters
(D) plastic trap adapters

40. Grease interceptors shall at all times be _____ for inspection, cleaning and removal of grease.
(A) easily accessible
(B) labeled with instructions
(C) documented by the manufacturer
(D) open

41. What is the proper size of a gravity grease interceptor serving a commercial kitchen containing the following?
2 - Commercial sinks with food waste
1 - Commercial food waste disposer
5 - Floor sinks receiving indirect waste flow rates of 8, 11.7, 31, 34, and 41 gallons per minute
(A) 1000-gallon
(B) 500-gallon
(C) 750-gallon
(D) 1250-gallon

42. What is the proper size of a hydromechanical grease interceptor (in gpm) with one minute drainage period and a flow control device serving a commercial kitchen containing the following?
2 - Commercial sinks with food waste
1 - Commercial food waste disposer
5 - Floor sinks receiving indirect waste flow rates of 8, 11.7, 31, 34, and 41 gallons per minute
(A) 20
(B) 75
(C) 150
(D) 250

43. When serving a “Food Court” with multiple business establishments, a grease interceptor may serve up to _____ business(es)?
(A) 1
(B) 3
(C) 4
(D) 6

CHAPTER 11
STORM DRAINAGE

The design, installation, and material for storm drain systems is regulated by the requirements of this chapter. The design of the storm drain system is determined by several factors, including the maximum rainfall in inches per hour, the area of the roof in square feet, and the slope for horizontal piping. The material for the storm drain system is determined by the location in which it will be used: inside of the building, outside of the building, or below ground. Strainers are required to be of an approved design and properly installed on all inlet openings, primary as well as secondary.

Because of the volume of rainwater passing through the storm drain system, testing to ensure a tight system is very important. The procedure for conducting this test is given in Section 1107.0.

1. Storm water shall not be drained into sewers intended for _____.
   (A) a combined sewer system
   (B) sanitary drainage only
   (C) a storm drain system
   (D) a storm sewer system

2. Subsoil drains shall be piped to a _____.
   (A) storm drain
   (B) street curb
   (C) alley
   (D) all of the above

3. The perimeter of buildings having basements, cellars, or crawl spaces or floors below grade shall be provided with _____.
   (A) backwater valves
   (B) seepage pits
   (C) subsoil drains
   (D) leech lines

4. All rainwater sumps serving "public use" occupancy buildings shall be provided with _____.
   (A) single pumps
   (B) dual pumps
   (C) separate sumps
   (D) overflow tanks

5. Roof drains, gutters, vertical conductors or leaders, and horizontal storm drains for primary drainage shall be sized based on _____.
   (A) 30-minute 30-year storm
   (B) 3 inches per hour
   (C) 60-minute 100-year return period
   (D) 6 inches per hour

6. The overflow level of the secondary drain inlet above the primary drain inlet shall be not less than _____ inch(es) above the roof surface.
   (A) 1
   (B) 2
   (C) 6
   (D) 12

7. Leaders and storm drains, where connected to a combined sewer, shall be _____.
   (A) trapped
   (B) identified
   (C) crown vented
   (D) oversized

8. An approved material for conductors installed above ground level in buildings is _____.
   (A) aluminum sheet metal
   (B) ABS Schedule 40
   (C) PVC (Sewer and Drain)
   (D) CPVC

9. The sizing data for horizontal storm drain piping is based on the pipes flowing _____.
   (A) 1/2 full
   (B) 3/4 full
   (C) full
   (D) at 2 percent velocity

10. Roof drain strainers of the flat surface type may be installed on _____.
    (A) sun decks
    (B) any roof
    (C) secondary roof drains
    (D) its upper terminal

11. Connection between the roof and roof drains that pass through the roof and into the interior of the building shall be made watertight by the use of _____.
    (A) expansion joints
    (B) caulking compound
    (C) solvent cement
    (D) proper flashing material

12. Roof areas of 10,000 square feet or less with controlled flow roof drainage shall have no less than _____.
    (A) two drains
    (B) one drain
    (C) one primary drain
    (D) one secondary drain (overflow drain)

13. Storm drain systems shall be tested upon completion of the rough piping by using _____.
    (A) 10 foot head of water
    (B) 4.34 psi air
    (C) 4.34 inches mercury
    (D) smoke or peppermint oil

14. Schedule 40 plastic DWV storm drain systems shall not be tested by _____.
    (A) water
    (B) air
    (C) 10 foot head of water
    (D) water from low to high point

15. Calculations for the controlled flow roof drainage system shall be submitted along with plans to the _____.
    (A) architect
    (B) plumber
    (C) design engineer
    (D) Authority Having Jurisdiction

16. The minimum size of a drain leader serving a roof with a projected area of 4,613 square feet and a rainfall rate of 3 inches per hour is _____.
    (A) 2 inches
    (B) 3 inches
    (C) 4 inches
    (D) 5 inches

17. The minimum size of a horizontal rainwater pipe with a 1/8 inch per foot slope, serving a roof with a projected area of 10,700 square feet, and a rainfall rate of 2 inches per hour is _____.
    (A) 2 inches
    (B) 3 inches
    (C) 5 inches
    (D) 6 inches

18. The size of semi-circular gutters shall be based on the maximum projected roof area and _____.
    (A) Table 1105.1(1)
    (B) Table 1101.8
    (C) Table 1101.12
    (D) Table 1103.3

19. For the purposes of calculating roof area, one vertical wall projecting above a roof will _____.
    (A) add 50 percent of wall area to roof figures
    (B) add 35 percent of wall area to roof figures
    (C) add no additional area
    (D) subtract 35 percent of wall area from roof figures

20. Subsoil drain lines shall be _____.
    (A) open jointed
    (B) perforated pipe
    (C) close jointed
    (D) A or B apply

21. Leaders installed along alleyways, driveways, or other locations where they may be exposed to damage shall be _____.
(A) constructed from ferrous pipe
(B) installed with extra-heavy couplings
(C) provided with expansion joints
(D) Type DWV copper tube

22. A connection between the roof drain and the secondary (overflow) drain in a separate piping system shall _____.
(A) be connected at vertical piping only
(B) not exist, and the secondary drain shall discharge independently above grade and in a location observable by building occupants
(C) occur belowground outside of the building
(D) occur below the roof structure, provided that the overflow drain shall be the same size as the roof drain

23. ABS and PVC-DWV installations shall be _____.
(A) limited to structures not exceeding three floors above grade
(B) installed in accordance with Chapter 14 "Firestop Protection"
(C) limited to structures of combustible construction
(D) limited to structures of combustible construction not exceeding three floors above grade

24. Vertical piping shall be _____.
(A) round
(B) square
(C) rectangle
(D) all of the above

25. Flat deck drains may be installed on _____.
(A) sun decks
(B) parking decks
(C) any roof
(D) A and B apply

26. Traps, where installed for individual conductors, shall be _____ the horizontal drain to which they are connected.
(A) two pipe sizes larger than
(B) 1/2 the cross-sectional area of
(C) the same size as
(D) twice the cross-sectional area of

27. With a projected roof area of 18,500 square feet in an area that has a rainfall rate of 3 inches per hour, what minimum size of horizontal pipe is required at a 1/4" per foot slope?
(A) 3 inch
(B) 5 inch
(C) 6 inch
(D) 8 inch

28. In the case of a single family dwelling, storm water from roofs, courtyards and similar locations, shall discharge to _____.
(A) sanitary sewage system
(B) private sewage disposal system
(C) sump with pump discharge into the house drain
(D) flat areas outside the building so long as the storm water flows away from the building

29. Siphonic roof drainage systems shall be designed by a(n) _____.
(A) installer
(B) Authority Having Jurisdiction
(C) registered design professional
(D) general contractor

CHAPTER 12
FUEL GAS PIPING

All plumbing fixtures in homes, such as kitchen sinks, bathtubs, lavatories, and laundry sinks shall be provided with hot running water. Energy to heat this water may come from several sources, the most common being fuel gas and electricity. When fuel gas is used, the piping installation is subject to the requirements of the Uniform Plumbing Code.

The regulations for fuel-gas piping systems include material, installation, location, piping supports, testing procedures, pipe sizing methods, location of shutoff valves, inspection procedures, permit and plan requirements, gas meter locations, and appliance connectors.

Piping outside of buildings may be above or below ground, whereas inside of buildings piping is not permitted below ground unless special provisions of the code are followed.

Gas piping systems are required to be tested after all portions that are to be covered or concealed and before any fixture, appliance, or shutoff valve has been attached thereto. One obvious reason is that the gas piping system might be damaged during the final stages of construction, and it is imperative that these damages be located so as to avoid a hazardous situation resulting from leaking gas.

1. The _____ is hereby authorized to disconnect gas piping or appliances or both that shall be found not to be in accordance with the requirements of this code or are found defective and in such condition as to endanger life or property.
   (A) homeowner and the architect
   (B) plumber and the architect
   (C) Authority Having Jurisdiction or the gas supplier
   (D) homeowner

2. The location of the point of delivery shall be acceptable to the _____.
   (A) Authority Having Jurisdiction
   (B) serving gas supplier
   (C) building owner
   (D) building occupant

3. Plastic pipe for fuel gas shall be used only _____.
   (A) outside underground
   (B) inside
   (C) both A and B apply
   (D) by a licensed plumber

4. The approximate number of threads to be cut on a piece of 1-inch metallic fuel gas pipe is _____.
   (A) 10
   (B) 11
   (C) 12
   (D) 13

5. When nonferrous pipe for fuel gas is brazed, the brazing alloys shall not contain more than _____ percent phosphorus.
   (A) 0.02
   (B) 0.05
   (C) 0.08
   (D) 0.12

6. Horizontal 1/2 inch diameter steel fuel gas piping shall be supported at intervals not exceeding _____ feet.
   (A) 6
   (B) 8
   (C) 10
   (D) 12

7. Full-face flange gaskets shall be used with _____ flanges.
   (A) non-ferrous
   (B) steel
   (C) aluminum
   (D) both A and C apply

8. Lapped flanges shall be located accessible for inspection and used only ______.
   (A) inside walls
   (B) inside cabinets
   (C) below ground
   (D) above ground

9. To prevent traps, gas piping for other than dry gas conditions, shall be sloped not less than ______.
   (A) 1/8 inch in 10 feet
   (B) 1/8 inch in 15 feet
   (C) 1/4 inch in 10 feet
   (D) 1/4 inch in 15 feet

10. Underground gas piping systems shall be installed with a cover not less than ______ inches.
   (A) 6
   (B) 12
   (C) 18
   (D) 24

11. Where gas piping is installed underground beneath buildings, the piping shall be one of the following:
   (A) polyethylene
   (B) 20 mil gas protective coating
   (C) encased in an approved conduit
   (D) encased in cement

12. Concealed gas piping shall not be located in ______.
   (A) solid partitions
   (B) attics
   (C) the ground
   (D) floors

13. A pipe chase shall be ventilated to the outdoors and at ______.
   (A) the top
   (B) the bottom
   (C) the sides
   (D) intervals of 10 feet

14. The approximate demand of a freestanding domestic gas range in Btu's per hour is ______.
   (A) 65,000
   (B) 40,000
   (C) 30,000
   (D) 25,000

15. Gas line pressure regulators installed in the interior of the building without a vent shall be ______.
   (A) removed and installed on exterior applications only
   (B) installed in a concealed but accessible location
   (C) separately vented to the exterior of the building
   (D) installed at least 6 feet aboveground and accessible

16. Aluminum alloy tubing shall not be used ______.
   (A) for inside locations
   (B) on water heaters
   (C) in exterior locations or underground
   (D) for gas piping

17. The inside radius of metallic gas pipe bends shall not be less than ______ times the outside diameter of the pipe.
   (A) 4
   (B) 6
   (C) 8
   (D) 10

18. One location gas meters shall not be placed is ______.
   (A) at the side of the house
   (B) adjacent to a driveway
   (C) at the back of the house
   (D) at the front of the house

19. The medium for a gas piping pressure test shall be ______.
   (A) carbon monoxide
   (B) carbon dioxide
   (C) oxygen
   (D) methane

20. Where external damage to underground gas piping is likely to occur, the cover shall be not less than _____ inches.
(A) 6
(B) 12
(C) 18
(D) 24

21. An accessible gas shutoff valve shall be provided _____ for each gas pressure regulator.
(A) upstream
(B) downstream
(C) individually
(D) separately

22. The gas pressure regulator shall be accessible for _____.
(A) inspection
(B) installation
(C) replacement
(D) servicing

23. Gas meters shall be protected against _____.
(A) positive pressure
(B) reduced pressure
(C) back pressure
(D) initial pressure

24. Appliance shutoff valves and convenience outlets shall serve a single appliance and shall be installed _____ of the appliance it serves.
(A) within 8 feet
(B) within 6 feet
(C) upstream of the connector
(D) both B and C

25. A _____ shall be provided downstream from the shutoff valve to permit removal of appliance controls.
(A) flanged connection
(B) quick disconnect
(C) A or D
(D) union

26. Shutoff valves shall be permitted to be accessibly located _____.
(A) under fixed appliances
(B) encased in a wall
(C) inside wall heaters or furnaces
(D) in non-accessible locations

27. Which of the following is not part of a rough gas piping inspection?
(A) Size of gas pipe
(B) Proper material
(C) Pressure test
(D) Installation meets the requirements of the code

28. "Rough Piping Inspection" of gas piping shall be made _____.
(A) after all gas piping authorized by the permit has been installed
(B) before all gas piping has been covered or concealed
(C) before any fixture or appliances has been attached thereto
(D) all of the above

29. If, upon final piping inspection, the installation is found to be in accordance with the provisions of this code, _____.
(A) no certificate of inspection is issued because there is no rejection to be noted
(B) a certificate of inspection shall be permitted to be issued by the Authority Having Jurisdiction
(C) a certificate of inspection may be issued by the Authority Having Jurisdiction
(D) the permit holder must request a certificate of inspection

30. It shall be unlawful to remove or disconnect any gas piping without _____.
    (A) obtaining a permit from the Authority Having Jurisdiction
    (B) capping or plugging the gas pipe outlet with a screw joint fitting
    (C) notifying the Authority Having Jurisdiction that the appliance or piping was removed and call for inspection
    (D) both A and C apply

31. Upon completion of the installation, alteration or repair of any gas piping and prior to the use thereof _____.
    (A) the piping shall be painted to prevent corrosion
    (B) the Authority Having Jurisdiction shall be notified that such piping is ready for inspection
    (C) appliances shall be connected and fired to be certain there are no leaks
    (D) the gas shall be turned on to detect the smell of gas if pipes are leaking

32. Leaks in gas piping shall be located _____.
    (A) with an approved gas detector, a noncorrosive leak detection fluid or other approved method
    (B) with a "presto-lite" flame
    (C) with water
    (D) by increasing the air pressure during the test until the leaks are heard as a whistle

33. Corrugated stainless tubing shall be listed in accordance with _____.
    (A) ASTM B 241
    (B) ASTM B 88
    (C) NFPA 58
    (D) CSA LC-1

34. Where leakage or other defects are located, the affected portion of the piping system shall be _____ and retested.
    (A) repaired
    (B) replaced
    (C) repaired by a certified welder
    (D) A and B apply

35. Liquefied petroleum gas facilities shall be in accordance with _____.
    (A) ASME B16.1
    (B) ASME B16.20
    (C) NFPA 58
    (D) ASTM D2513

36. When a flanged joint is separated, the gasket shall be _____.
    (A) repaired
    (B) replaced
    (C) repaired by the manufacturer's instructions
    (D) ignored and re-used

37. Pressure regulators shall be protected against _____.
    (A) physical damage
    (B) excessive heat
    (C) frost
    (D) corrosion

38. A gas appliance regulator vent shall terminate at least _____ feet from the ignition source.
    (A) 2
    (B) 3
    (C) 6
    (D) 10

39. The size of the supply piping outlet for a gas appliance shall not be less than _____.
    (A) 1/2 inch
    (B) 3/4 inch
    (C) 1 inch
    (D) as determined by the gas table in the code

40. The size of a piping outlet for a mobile home shall be not less than _____.
(A) 1/2 inch
(B) 3/4 inch
(C) 1 inch
(D) as determined by the gas tables in the code

41. Which material shall not be used for gas piping?
(A) cast-iron
(B) steel
(C) wrought-iron
(D) plastic

42. The hourly volume of gas required at each piping outlet shall be taken as not less than _____.
(A) the average use of the appliance
(B) the maximum hourly rating specified by the manufacturer
(C) the maximum hourly rating specified in the Plumbing Code
(D) specified by this gas supplier

43. To obtain the cubic feet per hour of fuel gas required for any gas appliance _____.
(A) divide the Btu input of the appliance by the average Btu heating value per cubic foot of gas
(B) divide the Btu input of the appliance by 1,000
(C) contact the Authority Having Jurisdiction or the local inspector
(D) obtain the manufacturer's specifications

44. Aluminum alloy pipe shall comply with _____.
(A) ASTM B210
(B) ASTM B241
(C) ASTM B310
(D) A or B Apply

45. When using an electrically continuous wire as the tracer for underground plastic gas piping, the wire is required to be buried with the pipe to facilitate locating and be not less than _____.
(A) 14 AWG
(B) 16 AWG
(C) 18 AWG
(D) 20 AWG

46. Indoor gas hose connectors shall be used to connect laboratory, shop, and ironing appliances or equipment requiring mobility during operation. The connector shall be of minimum length and shall not exceed _____ feet.
(A) 3
(B) 5
(C) 6
(D) 10

47. Plastic pipe and fittings used to connect regulator vents to remote vent terminations shall be _____.
(A) PVC
(B) CPVC
(C) PE
(D) PP

48. Where automatic excess flow valves are installed, they shall be listed, sized, and installed in accordance with the _____.
(A) NFPA standards
(B) job site specifications
(C) local administrative authorities
(D) manufacturer's instruction

49. If two or more gas outlets are present on a research laboratory space, what is required?
(A) gas leak detector system
(B) explosion resistant electrical fixtures
(C) a single shutoff valve
(D) All gas lines must be clearly labeled.

50. CSST gas piping systems shall be bonded to the electrical service grounding electrode system. The bonding jumper shall be not smaller than _____ copper wire or equivalent.
(A) 6 AWG
(B) 8 AWG
(C) 10 AWG
(D) 12 AWG

51. Except for CSST, gas piping with a service pressure of 1.58 psi shall be tested at a minimum of _____ psi for no less than _____ minutes.
(A) 3, 10
(B) 10, 15
(C) 30, 30
(D) 60, 30

52. Where the serving gas supplier delivers gas at a pressure greater than 2 psi for piping systems serving appliances designed to operate at a gas pressure of 14 inches of water column or less a _____ shall be installed.
(A) pressure regulator
(B) shut off valve
(C) overpressure protection device
(D) sediment trap

53. Where piping systems serving appliances designed to operate with a gas supply pressure of 14 inches water column or less are required to be equipped with overpressure protection, each device shall be adjusted to limit the gas pressure to each appliance to _____ psi or less upon failure of the line pressure regulator.
(A) 0.5
(B) 2
(C) 1
(D) 1.5

54. Where piping systems serving appliances designed to operate with a gas supply pressure greater than 14 inches water column are required to be equipped with overpressure protection, each device shall be adjusted to limit gas pressure to each appliance to _____.
(A) 14 inches water column
(B) 2 psi
(C) manufacturer's installation instructions
(D) 5 psi

55. Gas pipe and tubing installed underground shall be protected against corrosion with one or more of the following _____.
(A) corrosion-resistant material
(B) factory-applied electrically insulating coating
(C) cathodic protection
(D) any of the above

56. When protecting buried steel gas pipe against corrosion, _____ cathodic protection may be used.
(A) sacrificial anodes
(B) impressed current
(C) electrically insulated
(D) both A & C

57. Cathodic protection systems for buried steel gas piping, that have failed the mandatory performance testing shall be repaired not more than _____ days after the date of the failed testing.
(A) 120
(B) 60
(C) 90
(D) 180

58. Protective devices shall include but are not limited to the following:
(A) air-actuated positive shut-off pressure regulator
(B) check valve
(C) three-way valves
(D) all of the above

CHAPTER 13

HEALTH CARE FACILITIES AND MEDICAL GAS AND MEDICAL VACUUM SYSTEMS

The purpose of this chapter is to provide minimum requirements for special fixtures and systems installed in health care facilities and the installation, testing, and certification of medical gas and medical vacuum systems from the point of supply to the user outlets or inlets. Medical gas and medical vacuum systems shall be designed and installed in accordance with the requirements of this chapter and the installation requirements of this code. Systems shall be designed to provide for an adequate and safe supply of medical gas and vacuum from the point of supply to the user outlet or inlet. Installation and testing of medical gas and vacuum systems shall be inspected by the Authority Having Jurisdiction.

1. Prior to any medical gas system being placed in service, certification tests verified and attested to by the _____ shall be performed.
   (A) Authority Having Jurisdiction
   (B) design engineer
   (C) plumbing inspector
   (D) certification agency

2. O_2 is the symbol for the _____.
   (A) nitrous oxide
   (B) nitrogen
   (C) oxygen
   (D) helium

3. Tubing for medical gas systems shall be hard-drawn seamless copper medical gas tube _____.
   (A) Type L or K
   (B) Type L or M
   (C) Type K or M
   (D) Type DWV medical

4. Copper to copper joints shall be _____.
   (A) soldered
   (B) flared
   (C) brazed
   (D) silver soldered

5. Mains and branches in medical gas piping systems shall be not less than _____.
   (A) 1/4 inch
   (B) 1/2 inch
   (C) 3/8 inch
   (D) 1 inch

6. The minimum static pressure for nitrous oxide (N_2O) is _____.
   (A) 50 psig
   (B) 75 psig
   (C) 100 psig
   (D) 200 psig

7. Which of the following is not part of the certification test?
   (A) checking labels of the control valves
   (B) testing of the alarm systems
   (C) testing for cross-connections
   (D) pressure-testing tubing by dry nitrogen

8. The final vacuum test of a medical gas system is for a _____ duration.
   (A) 30 minute
   (B) 2 hour
   (C) 24 hour
   (D) 15 minute

9. The initial test of medical gas piping requires a minimum test pressure of _____ psi.
   (A) 60
   (B) 100
   (C) 50
   (D) 150

10. Medical-surgical vacuum sources shall consist of ______ vacuum pump(s).
    (A) two or more
    (B) one
    (C) four parallel piped
    (D) three 1,250 cfm

11. The intake on a compressor pump shall be piped to the outside with the end ______.
    (A) turned down and screened
    (B) 6 inches or more above ground
    (C) 50 feet from doors or windows
    (D) 40 feet from the air intake

12. Vacuum pump exhaust termination shall be at least ______ feet from any door, window, air intake, or other openings in buildings.
    (A) 30
    (B) 100
    (C) 25
    (D) 50

13. Where positive-pressure gas piping system operate at a higher pressure than the gauge pressure in Table 1305.1 it shall be labeled with the ______ of the gas.
    (A) name
    (B) pressure
    (C) volume
    (D) name and pressure

14. The standard designation colors for marking pipe carrying oxygen is ______.
    (A) green/white or white/green
    (B) green/black or black/green
    (C) blue/white
    (D) black/white

15. Which of the following is an approved method for labeling medical gas systems?
    (A) plastic color-coded tags
    (B) metal tags
    (C) adhesive markers
    (D) thermal printed DYMO label

16. Station outlets and inlets shall be installed in strict accordance with the ______.
    (A) designer's recommendations
    (B) hospital's regulations
    (C) standard installation procedures
    (D) manufacturer's instructions

17. Medical air compressors shall be installed in a well-lit, ventilated, and clean location and shall be ______.
    (A) accessible
    (B) sound-rated
    (C) located in a med-gas storage area
    (D) operated simultaneously

18. Piping and drain traps in psychiatric patient rooms shall be ______.
    (A) exposed
    (B) concealed
    (C) vandal-proof
    (D) conventional

19. New or replacement shutoff valves shall be of any type but shall also be ______.
    (A) 1/4 turn to off
    (B) 1/2 turn to off
    (C) 2 psi
    (D) 5 psi

20. An indirect waste pipe for a bed pan steamer shall have a trap seal of not less than ______.
    (A) 1 inch
    (B) 2 inches
    (C) 3 inches
    (D) 4 inches

21. The minimum flow rate for nitrogen is ______ per outlet.
    (A) 0.71 cfm
    (B) 1 scfm
    (C) 15 cfm
    (D) 2 scfm

22. Hard-drawn seamless copper medical gas tube shall be used where installed pressures exceed 185 psi and shall be _____.
(A) Type L or K
(B) Type K
(C) Type M
(D) Type L

23. Recleaning of the interior surfaces of tube ends requires the use of _____.
(A) steel wool
(B) steel wool or sand cloth
(C) clean hot-water alkaline solution such as trisodium phosphate
(D) nonabrasive pads

24. Threaded joints in medical gas piping systems shall be _____.
(A) U.S. machine threads
(B) tapered pipe threads complying with ASME B1.20.1
(C) nontapered pipe threads complying with ASTM 20.1.2
(D) right or left threads complying with Federal Standard Z.B.1900

25. Joints shall be brazed within _____ after surfaces have been cleaned for brazing.
(A) 30 minutes
(B) 1 hour
(C) 4 hours
(D) 8 hours

26. Flux shall only be used when brazing _____.
(A) Type K tubing
(B) isolation or control valves
(C) pressure-regulating equipment
(D) dissimilar metals

27. The cut end of tubes shall be deburred using a sharp, clean _____.
(A) reamer
(B) pocket knife or similar tool
(C) deburring tool
(D) round metal file listed for copper tube use

28. Medical gas and medical vacuum risers shall not be located in _____.
(A) elevator shafts
(B) intensive care areas
(C) pipe shafts
(D) patient rooms

29. Zone valves shall be installed so that the closure of one zone valve _____.
(A) shuts off all zone valves
(B) does not affect other areas
(C) shuts down all vacuum systems
(D) shuts down all medical systems

30. Medical gas piping installed in kitchens is _____.
(A) permitted in this code
(B) prohibited in this code
(C) permitted if protected from physical damage
(D) permitted if ambient temperature does not exceed 130˚F

31. Hangers for copper tube in damp locations shall have a _____ and be sized for copper tube.
(A) plastic coat
(B) copper finish
(C) positive restraint to movement
(D) fiber or cloth insert

32. Dryers, after-coolers, separators, and receivers shall be equipped with _____.
(A) p-traps
(B) drum traps
(C) deep seal traps
(D) drains

33. Before attachment of system components, all piping shall be subjected to a test pressure not less than ______.
    (A) system operating pressure
    (B) system operating pressure plus 100 percent
    (C) 20 percent higher than operating pressure for 24 hours
    (D) 150 psi

34. After successful completion of initial pressure tests, vacuum distributing piping shall be subjected to a standing vacuum test for ______.
    (A) 36 hours
    (B) 48 hours
    (C) 24 hours
    (D) 12 hours

35. Ice makers/storage chests shall be located in nurses stations or similarly supervised areas to minimize ______.
    (A) theft
    (B) vandalism
    (C) contamination
    (D) misuse

36. Receptors for indirect waste from sterilizers or bedpan steamers shall be ______.
    (A) located within the same room
    (B) located in a common area
    (C) in a mechanical equipment room
    (D) accessible to the patient

37. When a sterilizer has provisions for a vapor vent, the vent shall be ______.
    (A) equipped with backflow protection
    (B) extended to the outdoors above the roof
    (C) connected to the sanitary system vents
    (D) piped to an indirect waste receptor

38. Water-operated aspirators shall be installed only when approved by the ______.
    (A) health agencies
    (B) Authority Having Jurisdiction
    (C) American Medical Association
    (D) listing agency

39. Potable water supply to the aspirator shall be protected by ______.
    (A) a vacuum breaker or equivalent backflow protection device
    (B) a vacuum pump
    (C) an airgap
    (D) backflow protection is not necessary

40. This chapter does not apply to ______.
    (A) compressed gas systems
    (B) installation requirements
    (C) portable compressed gas systems
    (D) oxygen compatibility

41. The symbol for nitrogen is ______.
    (A) N_2O
    (B) N_2
    (C) N_2O_2
    (D) N

42. System verification testing shall be performed by a party other than the ______.
    (A) in-house personnel
    (B) installing contractor
    (C) design engineer
    (D) facility authority

43. Prior to the brazing process for a pipeline all ambient air must be removed and ______.
    (A) the presence of nitrogen confirmed
    (B) a continual purge of argon be used
    (C) verified by an oxygen analyzer
    (D) a vacuum maintained

CHAPTER 14
FIRESTOP PROTECTION

These questions are designed to familiarize the user with firestop requirements for DWV and storm water systems in a building with fire-resistance-rated walls, partitions, floors, floor/ceiling assemblies, roof/ceiling assemblies, or shaft enclosures. It is important to know the type of plumbing material being installed and the construction type and rating of the assembly being penetrated so the correct type of firestop material or system can be applied.

1. In addition to meeting the requirements for firestop protection in this chapter the protection of the penetration must meet requirements in the _____.
   (A) Fire Code
   (B) Mechanical code
   (C) design engineers regulation manual
   (D) Building Code

2. F and T ratings are tested in accordance with _____.
   (A) ANSI standards
   (B) ASTM E119 or ASTM E814
   (C) UL 263 or UL 1479
   (D) B or C may apply

3. Where sleeves are installed for a fire-resistance-rated assembly, they _____.
   (A) shall be prohibited unless they meet the F and P rating
   (B) shall be securely fastened to the rated assembly
   (C) shall meet the requirements of the NFPA 5000 Building Code
   (D) may be installed if twice the diameter of the penetration

4. Where piping penetrates a rated assembly, _____ piping shall not connect to noncombustible piping unless it is capable of being demonstrated that the transition is in accordance with this chapter.
   (A) ferrous
   (B) non-ferrous
   (C) PEX-AL-PE
   (D) combustible

5. The Authority Having Jurisdiction shall determine the _____ of penetrations to be inspected.
   (A) size and quality
   (B) type, size, and quality
   (C) type, size, and quantity
   (D) type, size, and verification

6. Unshielded couplings shall not be used to connect _____ piping unless it can be demonstrated that the fire-resistive integrity of the penetration is maintained.
   (A) ferrous
   (B) combustible
   (C) cast-iron
   (D) noncombustible

7. The Authority Having Jurisdiction shall include _____ as part of the condition of satisfactory compliance with this chapter.
   (A) proof of certification of installers
   (B) proof of verification of installers
   (C) destructive inspection
   (D) a signed affidavit certified by the fire department

8. Prior to being _____, piping penetrations shall be inspected by the Authority Having Jurisdiction to verify compliance.
   (A) installed
   (B) tested
   (C) certified
   (D) concealed

9. The (inside) annular space between the sleeve and the penetrating item and the (outside) annular space between the sleeve and the fire-resistance-rated assembly shall be firestopped in accordance with this chapter.

   (A) True

   (B) False

CHAPTER 15
ALTERNATE WATER SOURCES FOR NONPOTABLE APPLICATIONS

This chapter covers the design, installation and construction of gray water systems for underground landscape irrigation. The layout, design, and system shall be determined by several requirements. Materials, tank, disposal field, and valves for a gray water system are determined by the location, drawing, specifications, and percolation test. A gray water system may be installed for a subsurface or subsoil landscape irrigation system.

The chapter also covers the installation, design, alteration, and repair of reclaimed water systems and on-site treated nonpotable water systems and their intended uses.

1. Gray water installations exceeding a maximum discharge capacity of _____ gallons per day shall be designed by a registered design professional.
   (A) 150 (C) 200
   (B) 175 (D) 250

2. The total estimated gray water discharge for each occupant in a single family home shall be _____ gpd.
   (A) 15 (C) 35
   (B) 25 (D) 40

3. All valves, including the three-way valve, shall be _____.
   (A) accessible (C) non corrosive
   (B) metal (D) plastic

4. A reclaimed water system shall not be connected to a _____.
   (A) building sewer (C) subsurface drain
   (B) house sewer (D) potable water system

5. For gray water surge tanks of 75 gallons or less, the distance or clearance to a building or structure may be reduced to _____ ft.
   (A) 0 (C) 10
   (B) 5 (D) 15

6. All valves shall be equipped with a locking feature except for _____.
   (A) fixture supply control valves (C) gate valves
   (B) fullway valves (D) main valves

7. The maximum downstream pressure from a pump serving a gray water irrigation field shall not exceed _____ psi.
   (A) 10 (C) 20
   (B) 15 (D) 30

8. Subsurface irrigation fields and mulch basins shall be sized as per _____.
   (A) Table 703.2 (C) the inlet size
   (B) Table 1504.2 (D) Table 1502.1

9. Gray water supply piping, including drip feeders, shall be not be less than _____ inches below finished grade and covered with mulch.
   (A) 2 (C) 10
   (B) 6 (D) 12

10. The distance between a septic tank and a subsurface irrigation field shall be _____ feet.
    (A) 5
    (B) 10
    (C) 12
    (D) 20

11. Where irrigation and disposal fields are installed in sloping ground, the minimum horizontal separation between the distribution system and ground surface shall be _____ feet.
    (A) 5
    (B) 15
    (C) 30
    (D) 50

12. Gray water systems shall be designed to distribute the total amount of estimated gray water on a _____.
    (A) twice daily basis
    (B) daily basis
    (C) hourly basis
    (D) yearly basis

13. Treated nonpotable water pipes shall be permitted to run in the same trench as potable water pipes with a _____ inch minimum vertical and horizontal separation where both pipe materials are approved for use within a building.
    (A) 6
    (B) 12
    (C) 18
    (D) 24

14. On-site treated nonpotable water piping sizing shall comply with _____.
    (A) Chapter 6
    (B) Table 1502.4
    (C) Table 1502.1
    (D) Section 1601.12

15. On-site treated water systems shall have a colored background and marking information in accordance with Section _____ of this code.
    (A) 601.3
    (B) 1502.9.1
    (C) 1503.7
    (D) 1602.10

16. Upon discovery of a cross-connection, the potable water system shall be chlorinated with _____ parts-per-million chlorine.
    (A) 50
    (B) 75
    (C) 100
    (D) 10

17. In calculating for gray water systems, the first bedroom is equal to _____ occupants.
    (A) 1
    (B) 2
    (C) 3
    (D) 4

18. The gray water discharge amount for a laundry is _____ gallons per day/occupant.
    (A) 5
    (B) 10
    (C) 15
    (D) 20

19. Surge tanks placed aboveground shall be placed _____.
    (A) on a 1" concrete slab
    (B) on a 2" concrete slab
    (C) on a 3" concrete slab
    (D) on flat ground

20. A filter permitting the passage of particulates no larger than _____ microns shall be provided for on-site treated nonpotable water supplied to water closets, urinals, trap primers, and drip irrigation system.
    (A) 25
    (B) 50
    (C) 75
    (D) 100

21. Three-inch piping conveying on-site treated nonpotable water shall have lettering with a minimum size of _____ inch.
    (A) 1/2
    (B) 3/4
    (C) 1
    (D) 1-1/4

22. Each room containing reclaimed(recycled) water and onsite treated water equipment shall have signage with visible lettering of _____ inch.
    (A) 1/2
    (B) 3/4
    (C) 1
    (D) 1-1/4

23. The maximum absorption capacity for sandy clay is _____ gallons per feet of irrigation/leaching area for a 24-hour period.
    (A) 1.1
    (B) 1.7
    (C) 2.5
    (D) 4.0

24. The discharge from a _____ is not used to calculate gray water discharge.
    (A) bath or shower
    (B) kitchen
    (C) lavatory
    (D) laundry

25. No excavation for an irrigation field, disposal field, or mulch basin shall extend within _____ feet vertical of the highest known seasonal groundwater level, nor to a depth where gray water contaminates the groundwater or surface water.
    (A) 2
    (B) 3
    (C) 5
    (D) 6

26. A gray water system discharges 2,100 gallons per day into a sandy clay irrigation field. The minimum size of the area shall be _____ square feet.
    (A) 170
    (B) 1,260
    (C) 2,142
    (D) 6,000

27. On-site treated nonpotable water system that supplies water to urinals, trap primers and water closets shall have a minimum of _____ psi.
    (A) 10
    (B) 15
    (C) 20
    (D) 25

28. A reclaimed water system supplies a high-rise building at a residual pressure of 135 psi. A (An) _____ shall be installed prior to the water outlets of the reclaimed water system.
    (A) expansion tank
    (B) reduced pressure principle assembly
    (C) water hammer arrestor
    (D) pressure reducing valve

CHAPTER 16
NONPOTABLE RAINWATER CATCHMENT SYSTEMS

This chapter covers the design, installation and construction of nonpotable rainwater catchment systems. The layout, design, and system shall be determined by several requirements which are covered in this chapter.

Material compatibility, controls, separation requirements, storage tanks, cross-connection tests, markings and signage, maintenance, and regular inspections are essential elements of the design, installation and long term safety and usefulness of these systems.

1. The provisions of chapter 16 shall apply to the installation, construction, alteration, and repair of _____ rainwater catchment systems.
   (A) nonpotable
   (B) potable
   (C) acid
   (D) stored potable

2. Rainwater catchment systems intended to supply uses such as water features shall be _____.
   (A) approved by the Authority Having Jurisdiction
   (B) approved by the health inspector
   (C) prohibited
   (D) approved by the Dept. of Natural Resources

3. Where a portion of a rainwater catchment system is installed within a building, a _____ test is required in accordance with Section 1605.3.
   (A) cross-connection
   (B) pressure
   (C) smoke
   (D) dye

4. Hose bibbs supplying rainwater shall be labeled with the words: _____
   (A) "CAUTION: NONPOTABLE RAINWATER, DO NOT DRINK."
   (B) "CAUTION:DO NOT DRINK."
   (C) "CAUTION: RAINWATER, DO NOT DRINK."
   (D) "NONPOTABLE RAINWATER WATER, DO NOT DRINK."

5. Rainwater catchment tanks shall be marked _____.
   (A) "CAUTION"
   (B) "DO NOT DRINK"
   (C) "NONPOTABLE RAINWATER"
   (D) "CAUTION: NONPOTABLE WATER, DO NOT DRINK"

6. Rainwater shall be collected from _____.
   (A) roof surfaces
   (B) vehicular parking surfaces
   (C) surface water runoff
   (D) bodies of standing water

7. No treatment is required for rainwater catchment systems used for above ground irrigation where the maximum storage capacity is _____ gallons.
   (A) 380
   (B) 360
   (C) 370
   (D) 390

8. Rainwater tank access openings exceeding _____ inches in diameter shall be secured to prevent tampering and unintended entry by either a lockable device or other approved method
   (A) 8
   (B) 10
   (C) 11
   (D) 12

9. Pumps supplying water to water closets, urinals, and trap primers shall be capable of delivering not less than _____ pounds-force per square inch (psi) residual pressure at the highest and most remote outlet served.
   (A) 15
   (B) 25
   (C) 40
   (D) 60

10. Where the water pressure in the rainwater supply system within the building exceeds _____ psi, a pressure reducing valve shall be installed in accordance with this code.
    (A) 60
    (B) 70
    (C) 75
    (D) 80

11. A filter permitting the passage of particulates not larger than _____ microns shall be provided for rainwater supplied to water closets, urinals, trap primers, and drip irrigation system.
    (A) 50
    (B) 75
    (C) 100
    (D) 25

12. Each equipment room containing nonpotable rainwater equipment shall have a sign posted with wording in _____ inch letters.
    (A) 1
    (B) 2
    (C) 3
    (D) 4

13. A sign shall be installed in restrooms in commercial, industrial, and institutional occupancies using nonpotable rainwater for water closets, urinals, or both. Each sign shall contain _____ inch letters of a highly visible color on a contrasting background.
    (A) 1/4
    (B) 1/2
    (C) 3/4
    (D) 1

14. During the cross-connection test, the minimum period the rainwater catchment system is to remain depressurized shall in no case be less than _____ minutes.
    (A) 15
    (B) 30
    (C) 45
    (D) 60

15. The potable water system after cross-connection corrections shall be chlorinated with _____ ppm chlorine for _____ hours.
    (A) 25, 6
    (B) 30, 8
    (C) 50, 24
    (D) 24, 48

16. Rainwater holding tank covers shall be capable of supporting an earth load of not less than _____ pounds per square foot (lb/ft^2) where the tank is designed for underground installation.
    (A) 100
    (B) 200
    (C) 300
    (D) 250

17. The manhole opening on below grade rainwater tanks shall be located not less than _____ inches above the surrounding grade.
    (A) 4
    (B) 3
    (C) 2
    (D) 1

18. The design and size of rainwater drains, gutters, conductors, and leaders shall comply with Chapter _____ of this code.
    (A) 3
    (B) 6
    (C) 7
    (D) 11

19. Where a vent is provided for a rainwater storage tank, the vent shall extend from the top of the tank and terminate not less than _____ above grade.
(A) 12 inches
(B) 6 inches
(C) 7 feet
(D) 10 feet

20. A person registered or licensed to perform plumbing design work is not required to design rainwater catchment systems used for irrigation with a maximum storage capacity of _____ gallons.
(A) 250
(B) 300
(C) 350
(D) 360

21. For car washing, a filter permitting the passage of particulates not larger than _____ microns shall be provided.
(A) 5
(B) 10
(C) 25
(D) 100

22. Devices and equipment used to treat rainwater to maintain minimum water quality shall be listed or labeled by a listing agency and approved for the intended application.
(A) True
(B) False

CHAPTER 17
REFERENCED STANDARDS

Standards are documents that provide specific requirements and guidelines for the manufacture of products. Their purpose is to ensure products have been tested in accordance with the performance requirements of the appropriate standard, listed or labeled by a listing agency, and is safe for the user. Chapter 3 states that all components installed in a plumbing system shall conform to approved applicable recognized standards.

Table 1701.1 lists standards referenced within the body of Code and are considered as part of the requirements of the Code. Table 1701.2 consists of installation and testing standards that are not referenced in other sections of this code but may aid in the user's ability to quickly find an applicable standard. The application of both sets of referenced standards are specified in Section 301.2.2 of this Code. The standards from Table 1701.2 shall be permitted only after they have been approved by the Authority Having Jurisdiction. See also Section 301.2.2.

There is a list of abbreviations for firms listed in Tables 1701.01 and 1701.2 following Table 1701.2.

1. The standards listed in Table 1701.1 are intended for use in the design, testing, and installation of materials, devices, appliances, and equipment regulated by this code.
   (A) True
   (B) False

2. The approved standards for gas storage water heaters are developed by _____.
   (A) CSA
   (B) ASTM
   (C) ASME
   (D) UL

3. "Piping, Ferrous" would include _____.
   (A) copper
   (B) clay
   (C) steel
   (D) plastic

4. A horizontal backwater valve with lifting device is found listed under _____.
   (A) AHAM DWI
   (B) ASTM F667
   (C) IAPMO IGC 305
   (D) NSF 61

5. ANSI stands for _____.
   (A) American National Standards Institute
   (B) American National Steel Institute
   (C) American National Service Institution
   (D) American National Standards International

6. In the application column of Table 1701.2, "Piping, Ferrous," the standard for stainless steel pipe is _____.
   (A) ASTM A 312/312M-2018a
   (B) ASME B36.19M-2018
   (C) none of the above
   (D) both A and B apply

7. Listed as "Fuel Gas, Appliances," draft hoods are constructed to standard _____.
   (A) CSA Z21.12b-1994 (R2015)
   (B) CSA Z21.13-90 (R2008)
   (C) ANSI Z21. 90
   (D) ANSI Z121.12-90

8. Professional qualifications standards for medical gas systems personnel are certified to _____.
   (A) ASSE Series 5000-2015
   (B) ASSE Series 6000-2021
   (C) NFPA 85-2011
   (D) NFPA 99-2021

9. Flexible water connectors [Application - Piping] are listed under _____.
   (A) ASME A112.18.6- 2017
   (B) ASME A1112.18.6.99(R04)
   (C) ASTM 112.8.6-2001
   (D) ASME A12.1.6-2000

10. ASME A112.19.1-2018 (Enameled Cast-Iron Plumbing Fixtures) is also known as _____.
   (A) ASME A112.19.19-2001
   (B) ASTM D2774-2012
   (C) CSA B45.0-2020 (R2008)
   (D) CSA B45.2-2018

11. In the "application" column of Table 1701.1 titled "Fixtures," the correct standard for Hydromassage Bathtub System would be _____.
   (A) ASME A112.19.7-2020
   (B) ASME A112.19.3-2008
   (C) CSA B45.10-2020
   (D) A or C apply

12. ASME B16.4-2016 in Table 1701.1 is classified under the application of _____.
   (A) joints
   (B) fuel gas
   (C) fittings
   (D) miscellaneous

13. ASME B16.15-2018 (Cast Copper Alloy Threaded Fittings) covers classes _____.
   (A) 125 and 200
   (B) 125 and 250
   (C) 200 and 225
   (D) 300 only

14. ASME B16.33-2012 (R2017) covers piping system valves up to 175 psig for _____.
   (A) steam
   (B) hot water
   (C) gas
   (D) oil

15. The application referenced for ASME BPVC Section IX-2021 is _____.
   (A) classification
   (B) certification
   (C) clarification
   (D) miscellaneous

16. The type of backflow protection device indicated by ASSE 1001-2017 is _____.
   (A) AVB
   (B) PVB
   (C) DCV
   (D) RPV

17. The type of application ASSE 1020-2020 is indicated to be used for is _____.
   (A) pressure reducing
   (B) reverse osmosis
   (C) backflow protection
   (D) valves

18. The standard for Hot Water Dispenser, Household Storage Type, Electrical [Appliance] is _____.
   (A) ASME 1023-1979
   (B) ANSI I023-1979
   (C) ASSE 1023-2020
   (D) ASTM 1023-1979

19. Under ASTM A74-2021, as applied to Cast Iron Soil Pipe and Fittings, the application would be for _____.
   (A) piping, non-ferrous
   (B) piping
   (C) piping, plastic
   (D) piping, copper

20. ASTM C425-2021 indicates joints for _____ piping.
   (A) galvanized steel
   (B) bituminous
   (C) fiberglass
   (D) vitrified clay

21. AWWA stands for _____.
    (A) American Waste Water Agency
    (B) American Water Works Agency
    (C) American Water Works Association
    (D) none of the above

22. CISPI 301- 2021 standard for Hubless Cast Iron Pipe and Fittings applies to _____ piping.
    (A) wrought iron
    (B) wrought steel
    (C) non-ferrous
    (D) ferrous

23. ASTM F877- 2020 applies to _____ piping.
    (A) polyethylene
    (B) crosslinked polyethylene
    (C) PE-AL-PE
    (D) PEX-AL-PEX

24. In the "application" column of Table 1701.2, IAPMO PS 23-2021, Dishwasher Drain Airgaps, the type listed in the column is _____.
    (A) bathrooms
    (B) showers
    (C) fixtures
    (D) backflow protection

25. IAPMO PS 65-2019a applies to _____.
    (A) dishwasher air gaps
    (B) plumbing fixture air gaps
    (C) air gaps for water conditioners
    (D) none of the above

26. IAPMO PS 66-2015 pertains to _____.
    (A) twin fittings
    (B) electric components
    (C) dielectric fittings
    (D) grease traps

GENERAL EXAMINATIONS

The following examinations are comprised of questions about the Uniform Plumbing Code in general. There are three examinations designed to test your knowledge of the entire code. If you diligently studied the preceding specialty examinations, then you will probably miss very few of the questions in the general examinations.

The recommended procedure is to take General Examination #1 first. Review any question you missed by referring back to its specialty section in the preceding examinations. When you have completed this, take General Examination #2 and repeat the same procedure. By the time you take General Examination #3, you should have a good understanding of the code and its regulations. A good indication of this will be achieving a perfect score in General Examination #3.

GENERAL EXAMINATION #1

1. Building sewers shall be tested by completely filling the sewer with water from the _____.
   (A) lowest to the highest point
   (B) lowest to at least a 5-foot head
   (C) lowest to at least a 10-foot head
   (D) lowest to at least ground level

2. Except for excluded plastic piping systems, hot and cold water piping may be tested with _____.
   (A) 40 psi air or 150 psi water for 10 minutes
   (B) 35 psi air for at least 15 minutes
   (C) 50 psi air or the working pressure of the water in the system for at least 15 minutes
   (D) 100 psi air for at least 10 minutes

3. The required clearance in front of a 2-inch cleanout is _____ inches.
   (A) 8
   (B) 12
   (C) 18
   (D) 24

4. A stack, as defined in the code, is _____.
   (A) the main vent
   (B) any vertical vent
   (C) the vertical main of soil, waste, or vent piping that extends one or more stories
   (D) the horizontal main of soil, waste, or vent stack

5. A gas appliance regulator vent shall terminate at least _____ feet from the ignition source.
   (A) 2
   (B) 3
   (C) 6
   (D) 10

6. Aboveground schedule 40 PVC and ABS DWV plastic piping installed horizontally shall be supported at intervals of not to exceed _____ feet.
   (A) 4
   (B) 6
   (C) 10
   (D) 12

7. When sizing a copper fuel-gas piping system that has an inlet pressure of less than 2 psi with a pressure drop of 0.3 inch w.c., the maximum cubic feet per hour of gas flow allowed through a 5/8-inch OD. ACR tube 75 feet in developed length is _____ CFH.
   (A) 13
   (B) 27
   (C) 48
   (D) 68

8. A Type B-W gas vent shall terminate at least _____ feet in vertical height above the bottom of the wall furnace.
   (A) 4
   (B) 5
   (C) 10
   (D) 12

9. The maximum horizontal distance of a trap arm is measured from the inner edge of the vent to the _____.
   (A) weir of the trap
   (B) dip
   (C) inlet
   (D) outlet

10. The maximum number of lavatories on a 2 inch horizontal waste line is _____.
    (A) 8
    (B) 14
    (C) 16
    (D) 24

11. Sleeves shall be provided to protect piping through _____ and _____ walls.
    (A) floors, concrete
    (B) concrete, masonry
    (C) wood, concrete
    (D) none of the above

12. PEX piping one inch and smaller installed horizontally shall be supported at intervals not exceeding _____ inches.
    (A) 24
    (B) 32
    (C) 36
    (D) 48

13. The minimum common horizontal waste pipe serving a bathtub, a lavatory, and a kitchen sink shall be at least _____ inches.
    (A) 1-1/2
    (B) 2
    (C) 2-1/2
    (D) 3

14. The required clearance in front of a 3-inch cleanout is _____ inches.
    (A) 6
    (B) 12
    (C) 18
    (D) 24

15. A refrigerator used to store food and that requires drainage shall be drained by means of _____.
    (A) an indirect waste pipe
    (B) a direct connection
    (C) both A and B
    (D) no drain required

16. Sheet metal, constituting a part of any vent connector, shall be at least _____ inches.
    (A) 0.0304
    (B) 0.0450
    (C) 0.0500
    (D) 0.0505

17. One material not approved for the installation of an external trap for a urinal is _____.
    (A) cast iron
    (B) cast brass
    (C) drawn-copper alloy tubing traps
    (D) ABS

18. Cast-iron piping joined with lead and oakum installed horizontally in 10-foot lengths, shall be supported at intervals of not more than _____ feet.
    (A) 5
    (B) 10
    (C) 12
    (D) 15

19. A drain connecting the compartments of a set of fixtures to a trap is called a (an) _____.
    (A) continuous waste
    (B) combination waste and vent
    (C) indirect waste
    (D) special waste

20. A device to prevent backflow into the potable water system is called a _____.
    (A) backflow connection
    (B) backflow preventer
    (C) back-siphonage
    (D) siphon leg

21. Copper tube and fittings for drainage and vent piping installed aboveground shall be at least _____.
    (A) Type L
    (B) Type K
    (C) Type M
    (D) Type DWV

22. A _____ shall be permitted to be installed with indirect waste.
(A) water closet
(B) domestic kitchen sink
(C) lavatory
(D) drinking fountain

23. Vitrified clay pipe and fittings are _____.
(A) approved for use as a building drain and shall be kept not less than 12 inches below ground
(B) approved for use as a building drain inside of buildings and above ground
(C) unrestricted in use for drain lines
(D) unrestricted in use for vent lines

24. Cast-iron soil pipe shall not be _____.
(A) used for drain lines
(B) used below ground
(C) painted
(D) threaded

25. All trenches deeper than the footing of any building or structure and paralleling the same must be at least _____ from the bottom exterior edge of the footing or as approved.
(A) 45 degrees
(B) 4 feet
(C) 2 feet wide
(D) 2 feet deep

26. A type of sewer or drainage pipe that shall not be installed under or within 2 feet of any building or structure is _____.
(A) PVC (Sewer and Drain)
(B) cast iron
(C) ABS
(D) copper

27. Cast-iron soil pipe stacks shall be supported at _____.
(A) their base and each floor
(B) every other story
(C) intervals of 10 feet
(D) intervals of 5 feet

28. Water pipes shall not be run or laid in the same trench with _____ unless specific conditions are met.
(A) PVC (Sewer and Drain)
(B) cast-iron building sewer piping
(C) piping for a lawn sprinkler
(D) any other water pipe

29. Notification for inspection of work authorized by the permit shall be made _____.
(A) on the morning of the day inspection is requested
(B) not less than 24 hours before the work is to be inspected
(C) orally to the inspector
(D) in writing to the owner

30. _____ water piping systems shall be permitted to be tested with air.
(A) PVC
(B) PEX
(C) CPVC
(D) PB

31. The minimum size waste pipe for a domestic kitchen sink is _____ inches.
(A) 1-1/2
(B) 2
(C) 3
(D) 4

32. The minimum size trap and trap arm for a kitchen sink is _____ inches.
(A) 1-1/2
(B) 2
(C) 3
(D) 4

33. Horizontal drainage lines connecting with other horizontal drainage lines shall enter through _____.
    (A) 45-degree wye branches
    (B) sanitary tee branches
    (C) tapped tee branches
    (D) side inlet quarter bends

34. In each run of horizontal drain piping, the distance between cleanouts shall not exceed _____.
    (A) 50 feet
    (B) 75 feet
    (C) 100 feet
    (D) 300 feet

35. Three-inch horizontal drainage piping shall be run in practical alignment and at a uniform slope of not less than _____ inch per foot.
    (A) 1/16
    (B) 1/8
    (C) 1/4
    (D) 1/2

36. Where a fixture is installed on a floor level below the elevation of the next upstream manhole, the fixtures shall be protected from backflow of sewage by installing _____.
    (A) an inverted wye
    (B) a gate valve
    (C) a backwater valve
    (D) a globe valve

37. A sewage ejector discharge line shall be provided with a _____.
    (A) backwater valve only
    (B) gate valve only
    (C) back water valve and gate valve
    (D) swing check valve and a globe valve
    (E) either C or D apply

38. The maximum number of water closets on a 3 inch stack is _____.
    (A) two
    (B) four
    (C) five
    (D) six

39. Unless the pipe size is increased for its entire length, the maximum length for a 1-1/2 inch vent pipe is _____ feet.
    (A) 20
    (B) 30
    (C) 50
    (D) 60

40. The maximum fixture units allowed on a 4-inch horizontal drain line with a 2 percent slope is _____.
    (A) 180
    (B) 216
    (C) 256
    (D) 300

41. A 2 inch waste stack will serve a maximum of _____.
    (A) 24 lavatories
    (B) eight kitchen sinks
    (C) 10 bathtubs
    (D) two water closets

42. The minimum size of a building sewer shall be determined on the basis of the total number of fixture units drained by such sewer in accordance with _____.
    (A) maximum fixture load only
    (B) Table 717.1 fixture unit loading
    (C) the same table(s) as the building drain
    (D) none of the above

43. A fixture vent pipe increased one pipe size its entire length _____.
    (A) may exceed the maximum horizontal length otherwise limited by code
    (B) will increase the venting efficiency
    (C) will reduce the cost of installation
    (D) will be a better workmanship job

44. A plumbing vent shall terminate not less than _____.
(A) 3 feet from an air intake
(B) 10 feet from an air intake
(C) 1 foot above the roof
(D) 5 feet above a window

45. Each drainage stack that extends 10 or more stories above the building drain shall be served by _____.
(A) cleanouts at each floor
(B) a parallel vent stack
(C) extra-heavy fittings
(D) straps every 5 feet

46. A domestic dishwasher drain shall first connect to _____.
(A) an approved airgap fitting
(B) a double-check valve
(C) a backflow valve
(D) a hartford loop

47. An indirect waste pipe 12 feet long requires _____.
(A) a trap and vent
(B) a vent only
(C) a trap only
(D) no trap or vent

48. Drinking fountains may be installed with _____.
(A) indirect wastes
(B) bell traps
(C) steel pipe
(D) yoke venting

49. Chemical or industrial liquid wastes that are likely to damage the public sanitary sewer system shall _____.
(A) be drained into a holding tank
(B) not be used under any condition
(C) be pretreated to render them innocuous
(D) be drained into a storm drain

50. The vent for a kitchen sink also serves as a drain for a laundry sink. The size of the wet-vented section shall not be less than _____ inches.
(A) 1-1/2
(B) 2
(C) 3
(D) 4

51. Special venting for island fixtures is used for _____.
(A) bathtubs and similar equipment
(B) showers and similar equipment
(C) island sinks and similar equipment
(D) water closets and similar equipment

52. A branch line serving a trap in a combination waste and vent system is required to be separately vented when the length of the branch exceeds _____ feet.
(A) 5
(B) 10
(C) 15
(D) 25

53. A vent in a combination waste and vent system shall not be less than one-half the _____.
(A) inside cross-sectional area of the drain pipe served
(B) diameter of the drain pipe served
(C) area of the house sewer
(D) area of the public sewer

54. One trap may serve a set of not more than _____.
(A) three single-compartment sinks
(B) two double-compartment sinks
(C) two lavatories
(D) three showers

55. The vertical distance between a fixture outlet and the trap weir shall not exceed _____ inches.
(A) 12
(B) 18
(C) 24
(D) 36

56. The developed length of 1-1/2-inch trap arms shall not exceed _____.
(A) 2 feet, 8 inches
(B) 3 feet, 6 inches
(C) 5 feet
(D) 6 feet

57. The developed length between the closet flange of a water closet and its vent shall not exceed _____.
(A) 30 inches
(B) 4 feet
(C) 6 feet
(D) 15 feet

58. Fixture traps shall have a water seal of not less than _____ inches.
(A) 2
(B) 4
(C) 6
(D) 12

59. The total capacity in gallons of fixtures discharging into any such hydromechanical grease interceptor shall not exceed _____ times the certified gpm flow rate.
(A) 1-1/2
(B) 2
(C) 2-1/2
(D) 3

60. The minimum horizontal distance between a water supply well and the building sewer constructed of PVC (Sewer and Drain) pipe and fittings is _____ feet.
(A) 10
(B) 15
(C) 50
(D) 100

61. Three-quarter inch or 1-inch steel gas piping installed horizontally shall be supported at intervals not to exceed _____ feet.
(A) 4
(B) 5
(C) 6
(D) 8

62. Unions shall be installed in a water supply system within _____ inches of regulating equipment, water heating, conditioning tanks or similar equipment that require service by removal or replacement.
(A) 12
(B) 24
(C) 36
(D) 48

63. Wall-mounted water closet fixtures shall be securely bolted to _____.
(A) the vent piping
(B) the drain piping
(C) an approved carrier fitting
(D) a closet bend

64. Joints at the roof around pipes shall be made watertight by the use of _____.
(A) tar and paper
(B) approved flashings
(C) waterproof paint
(D) epoxy glue

65. Soil and drain pipes located above food-handling establishments shall be subjected to a standing water test of not less than _____ feet.
(A) 10
(B) 12
(C) 15
(D) 25

66. Concealed slip joint connections require a minimum ______ inch access panel without obstructions for inspection and repair.
(A) 8 x 12
(B) 10 x 12
(C) 12 x 12
(D) 18 x 18

67. Where exposed or accessible a fixture tailpiece shall be permitted to be of seamless drawn brass not less than _____.
    (A) 20 B&S gauge
    (B) 22 B&S gauge
    (C) 24 B&S gauge
    (D) 26 B&S gauge

68. Lavatories shall have a waste outlet and fixture tailpiece not less than _____ inches in diameter.
    (A) 1-1/4
    (B) 1-3/8
    (C) 1-1/2
    (D) 1

69. The distance from the center of a water closet to a side wall or obstruction shall not be closer than _____ inches.
    (A) 15
    (B) 30
    (C) 12
    (D) 24

70. No trap for any clothes washer standpipe receptor shall be installed _____.
    (A) 6 inches above the floor
    (B) below the floor
    (C) in a partition
    (D) 18 inches above the floor

71. When supplied, the depth of the finished dam or threshold for any shower, measured from the top of the drain, shall not be less than _____ inches.
    (A) 2
    (B) 4
    (C) 9
    (D) 12

72. A type B or L gas vent may have one offset of not more than 60 degrees from the _____.
    (A) horizontal
    (B) vertical
    (C) roof
    (D) heater

73. When using asphalt-impregnated felt as the lining material for built on-site shower receptors, it shall have not less than _____ with each succeeding layer hot-mopped to that below.
    (A) one layer of 15 pound felt
    (B) two layers of 15 pound felt
    (C) three layers of 15 pound felt
    (D) four layers of 15 pound felt

74. The critical level of the vacuum breaker serving a urinal above the highest part of the fixture shall be not less than _____ or the distance according to its listing.
    (A) 1 inch
    (B) 3 inches
    (C) 6 inches
    (D) 12 inches

75. Water closet tank ballcocks shall be installed with the vacuum breaker critical level at least _____.
    (A) 1 inch above the full opening of the overflow tube
    (B) 6 inches above the full opening of the overflow tube
    (C) 1 inch above the flush valve opening
    (D) 6 inches above the flush valve opening

76. Where static water pressure in the water supply piping is exceeding _____ psi, an approved-type pressure regulator preceded by an adequate strainer shall be installed.
    (A) 80
    (B) 100
    (C) 125
    (D) 150

77. A water system containing storage water-heating equipment will require a ______.
    (A) check valve
    (B) listed combination temperature and pressure relief valve
    (C) hot water return line
    (D) mixing valve

78. In addition to the appropriate use of fittings, changes in direction in copper tubing used for water distribution may be made with ______.
    (A) templates
    (B) heat
    (C) bends
    (D) rubber hammers

79. The terminal end of the drain for a pressure-relief valve shall not be ______.
    (A) painted
    (B) open
    (C) threaded
    (D) reamed

80. Cleanouts may be omitted on a horizontal drain line less than 5 feet except for ______.
    (A) sinks
    (B) showers
    (C) bathtubs
    (D) water closets

81. Because of structural conditions, a 4-inch house drain is installed with a slope of 1/8 inch per foot. This installation will result in a (an) ______.
    (A) separation of solids and liquid
    (B) increased scouring and cleansing action
    (C) reduction in fixture unit loading allowance
    (D) increased allowance in fixture unit loading

82. Each trap, except for traps within a (an) ______ are required to be self-cleaning.
    (A) interceptor
    (B) indirect waste system
    (C) food-handling establishment
    (D) sewage ejector

83. Underground gas piping systems shall be installed with a cover not less than ______ inches. Where external damage to the pipe or tubing from external forces is likely to result, the cover shall be not less than ______ inches.
    (A) 6, 12
    (B) 6, 18
    (C) 12, 18
    (D) 18, 24

84. In a common gas system serving a number of individual buildings, each building shall be equipped with a ______.
    (A) gate shutoff valve
    (B) globe shutoff valve
    (C) shutoff valve
    (D) lubricated shutoff valve

85. An appliance shutoff valve shall be within ______ feet of the appliance it serves.
    (A) 2
    (B) 3
    (C) 6
    (D) 10

86. The approximate gas input, in Btu per hour, for a typical freestanding domestic gas range is ______.
    (A) 3,000
    (B) 25,000
    (C) 40,000
    (D) 65,000

87. A method of locating leaks in a gas piping system shall be by applying _____.
    (A) a noncorrosive leak detection fluid
    (B) fire or acid to the pipe joints
    (C) a spark igniter to the exterior of the piping
    (D) potable water to the interior of the piping tubing

88. The minimum size of gas piping supply to a mobile home is _____.
    (A) 1/2 inch
    (B) 3/4 inch
    (C) 1 inch
    (D) 2 inches

89. The minimum clearance of listed water heaters to any combustible wall or partition is _____.
    (A) as close as practicable to the wall
    (B) specified in accordance with their listing and the manufacturer's installation instructions
    (C) 12 inches from the wall
    (D) 18 inches from the wall for servicing

90. Fuel-burning water heaters installed in a closet with a listed, gasketed door assembly and listed self-closing device may be located in a _____.
    (A) garage
    (B) kitchen
    (C) basement
    (D) bedroom

91. Burners and burner ignition devices of gas-fired water heaters (except for those listed as flammable vapor ignition resistant) in residential garages and in adjacent spaces that open to the garage shall be located not less than _____ inches above the floor level.
    (A) 6
    (B) 12
    (C) 18
    (D) 24

92. Water heater vents shall terminate above a roof surface at least _____ from a vertical wall.
    (A) 6 inches
    (B) 12 inches
    (C) 3 feet
    (D) 8 feet

93. The exit terminals of mechanical draft systems shall be _____.
    (A) not less than 7 feet above grade
    (B) in the same store
    (C) not over 10 feet from the vent
    (D) not over 20 feet from the vent

94. Equipment and appliances located on roofs of buildings more than _____ shall have an inside means of access to the roof.
    (A) 20 feet from the outside
    (B) 4 feet away from the wall
    (C) 15 feet in height
    (D) 12 feet in height

95. Single-wall metal vent pipe shall be constructed of _____ not less than 0.0304 inch thick.
    (A) copper sheet metal
    (B) cast iron
    (C) PVC Sch 40
    (D) galvanized, sheet steel

96. Single-wall metal pipe shall terminate at least 5 feet in vertical height above the highest connected appliance _____.
    (A) or with an approved cap on the roof
    (B) within a horizontal distance of 10 feet
    (C) draft hood
    (D) draft hood outlet or flue collar

GENERAL EXAMINATION #2

1. A plumbing permit expires if work authorized by such permit is not commenced within _____.
   (A) 60 days
   (B) 90 days
   (C) 120 days
   (D) 180 days

2. As long as such repairs do not require the replacement or rearrangement of valves, pipes or fittings, a plumbing permit is not required _____.
   (A) when work is done by the owner
   (B) for the clearing of stoppages
   (C) when the contract is less than $100.00
   (D) for removal and replacement of a water closet

3. The maximum fixture units on a 2 inch horizontal waste line is _____.
   (A) four
   (B) six
   (C) eight
   (D) five

4. The common vent pipe size for five domestic kitchen sinks is _____.
   (A) 1-1/2 inches
   (B) 2 inches
   (C) 3 inches
   (D) 4 inches

5. A refrigerator, not regularly classed as a plumbing fixture may discharge into a/an _____.
   (A) trap
   (B) laundry sink
   (C) bathtub
   (D) approved open receptor

6. Per Table 702.2, one fixture unit has a discharge capacity of _____.
   (A) up to 7-1/2 gpm
   (B) 8 to 15 gpm
   (C) 18 to 20 gpm
   (D) over 20 gpm

7. Gas appliances connected to a piping system shall have an accessible, approved manual shutoff valve within _____ feet of the equipment it serves.
   (A) 3
   (B) 4
   (C) 5
   (D) 6

8. The size of the supply piping outlet for a gas appliance shall not be less than _____ inch.
   (A) 1/4
   (B) 3/8
   (C) 1/2
   (D) 3/4

9. Except for direct vent appliances, a through-the-wall mechanical draft venting system shall terminate at least _____ feet above any forced air inlet located within 10 feet.
   (A) 2
   (B) 3
   (C) 4
   (D) 8

10. A fixture trap is vented to _____.
    (A) prevent the loss of a trap seal by siphonage or back pressure
    (B) add to the cost of installation
    (C) reduce noises in drain systems
    (D) prevent cross-connections

11. Except for excluded plastic piping, water piping systems shall be permitted to be tested with a (an) ______.
   (A) water pressure test of 10 psi
   (B) air pressure test of 25 psi
   (C) no test required
   (D) air pressure test of 50 psi

12. A valve that has the highest friction loss to water flow is the ______.
   (A) gate valve
   (B) angle valve
   (C) flush valve
   (D) globe valve

13. Gas piping shall be tested with ______.
   (A) water
   (B) oxygen
   (C) air
   (D) ether

14. ABS DWV pipe installed horizontally shall be supported at intervals not to exceed ______.
   (A) 2 feet
   (B) 3 feet
   (C) 4 feet
   (D) 5 feet

15. Three-inch horizontal drainage piping shall have a uniform slope of not less than ______.
   (A) 3/8 inch per foot
   (B) 1/16 inch per foot
   (C) 1/2 inch per foot
   (D) 1/4 inch per foot

16. In sizing gas piping, the length of the system is measured from the meter location to ______.
   (A) where the first branch takes off
   (B) the kitchen range, if located near the meter
   (C) the most remote outlet on the system
   (D) a wall heater, if located in a bathroom

17. Where supplementary gas for standby use is connected downstream from a meter, a ______ shall be installed.
   (A) device to prevent backflow
   (B) globe valve
   (C) gate valve
   (D) regulator

18. Indoor gas hose connectors shall be of minimum length and shall not exceed ______ feet.
   (A) 2
   (B) 3
   (C) 5
   (D) 6

19. Which material may be installed underground as part of a drain, waste and vent system?
   (A) galvanized wrought-iron pipe
   (B) cast-iron pipe
   (C) galvanized steel
   (D) aluminum pipe

20. With some exceptions the upper terminal of each horizontal drainage pipe shall be provided with ______.
   (A) valves
   (B) vent pipes
   (C) a cleanout
   (D) traps

21. The maximum number of bathtubs allowed on a 2-inch vent is ______.
   (A) 8
   (B) 12
   (C) 16
   (D) 20

22. Six bathtubs and six showers require a common vent of not less than ______ inches.
   (A) 1-1/2
   (B) 2
   (C) 3
   (D) 4

23. Where necessary, to insert fittings into a gas pipe that has been installed in a concealed location, of those listed the pipe shall be reconnected with _____.
    (A) unions
    (B) right and left couplings
    (C) swing joints
    (D) none of the above

24. A _____ is not a fitting approved for changes in direction of drainage flow.
    (A) wye
    (B) 60 degree wye
    (C) 1/8 bend
    (D) 45 degree tee

25. Waste and vent pipes may be tested with _____.
    (A) 5 foot head of water
    (B) a smoke test
    (C) 15 psi air
    (D) 10 foot head of water

26. Fuel gas piping pressure tests shall not be less than _____.
    (A) 10 pounds
    (B) 15 pounds
    (C) 20 pounds
    (D) 1-1/2 times the proposed maximum working pressure

27. Type B gas vents shall terminate not less than _____ feet above the highest connected appliance draft hood or flue collar.
    (A) 2
    (B) 3
    (C) 5
    (D) 10

28. No _____ or urinal shall be installed on combination waste and vent systems.
    (A) floor drain
    (B) water closet
    (C) shower drain
    (D) floor sink

29. Gas meters shall be located in _____ locations.
    (A) secure
    (B) readily accessible
    (C) accessible
    (D) unventilated

30. A water closet seat shall be sized for the _____.
    (A) water closet tank hardware
    (B) water closet bowl type
    (C) floor covering material
    (D) water supply type

31. The minimum size gas supply for a mobile home is _____.
    (A) 1-1/2 inches
    (B) 1 inch
    (C) 3/4 inch
    (D) 1/2 inch

32. A pressure regulator shall be installed when the static water pressure is exceeding _____.
    (A) 80 psi
    (B) 125 psi
    (C) 100 psi
    (D) 150 psi

33. An expansion tank is required when _____.
    (A) the pressure exceeds 150 psi
    (B) the water heater is in the basement
    (C) a backflow preventer is installed
    (D) the building is four stories high

34. A backwater valve is used _____.
    (A) to back water up the pipe
    (B) to protect the system from back-siphonage
    (C) to act as a bypass
    (D) to prevent backflow of sewage in a drainage system

35. All devices or assemblies installed in a potable water supply for protection against backflow shall be maintained in good working condition by _____.
    (A) the person having control of such devices
    (B) the plumber
    (C) the inspector
    (D) the landscaper

36. The approved fitting for connecting a horizontal drain line with a horizontal drain line would be a _____.
    (A) double tee
    (B) double wye
    (C) combination wye and 1/8 bend
    (D) sanitary tee

37. The maximum full pipe flow in gpm for a hydromechanical grease interceptor with a 60 gpm rating requires _____ gpm size of grease interceptor for a 2 minute drainage period.
    (A) 125
    (B) 75
    (C) 10
    (D) 35

38. After allowing for friction and other pressure losses, the residual water pressure shall be of not less than _____ psi.
    (A) 5
    (B) 10
    (C) 15
    (D) 20

39. Gas meters shall not be placed where they will be subjected to damage such as _____.
    (A) adjacent to a driveway
    (B) under a fire escape
    (C) in a well-lighted area
    (D) both A and B apply

40. The maximum fixture units on a 4 inch drainage stack are _____.
    (A) 256
    (B) 180
    (C) 300
    (D) 216

41. Corrugated vent connectors shall _____.
    (A) not be smaller than the listed appliance categorized vent diameter
    (B) be sized as one-half the flue collar diameter
    (C) be larger than the draft hood outlet diameter
    (D) sized using Table 510.2.1

42. The minimum dimension for an outdoor combustion air opening for a water heater shall not be less than _____ inches.
    (A) 2
    (B) 3
    (C) 6
    (D) 12

43. The developed length of 1-1/2 inch trap arms shall not exceed _____.
    (A) 4 feet
    (B) 5 feet
    (C) 6 feet
    (D) 3 feet, 6 inches

44. Valves used to control 2 or more openings for water distribution shall be _____.
(A) nonrising stem globe valve
(B) fullway gate valve
(C) heavy-duty globe valve
(D) ground key valve

45. A sewage ejector receiving the discharge of water closets or urinals shall have a discharge capacity of not less than _____ gallons per minute.
(A) 10
(B) 15
(C) 20
(D) 40

46. Using Appendix A, the pressure at the base of a column of water 10 feet high is approximately _____ psi.
(A) 0.43
(B) 4.3
(C) 10
(D) 43

47. The vent pipe serving a combination waste and vent system with a 4-inch drain pipe shall be at least _____ inches.
(A) 1-1/2
(B) 2
(C) 3
(D) 4

48. A 2-inch horizontal copper-tubing cold water line shall be supported at intervals of _____ feet.
(A) 12
(B) 10
(C) 8
(D) 6

49. The maximum length of an indirect waste pipe without a vent is less than _____ feet.
(A) 5
(B) 10
(C) 12
(D) 15

50. Traps are required for indirect waste pipes where the length exceeds _____ feet.
(A) 5
(B) 10
(C) 12
(D) 15

51. The maximum temperature of any water discharging under pressure directly into any part of a plumbing or drainage system is _____.
(A) 140°F
(B) 160°F
(C) 180°F
(D) 212°F

52. The maximum distance allowed for a 2-inch trap arm is _____.
(A) 2 feet
(B) 3 feet
(C) 3 feet, 6 inches
(D) 5 feet

53. The developed length between a closet flange of a water closet and its vent shall not exceed _____ feet.
(A) 4
(B) 6
(C) 8
(D) 10

54. Fixture traps shall have a liquid seal of not less than _____ inch(es).
(A) 1
(B) 2
(C) 3
(D) 4

55. Which of the following materials is prohibited by code for drainage and vent piping belowground?
(A) Type L copper tubing
(B) cast-iron pipe
(C) galvanized pipe
(D) ABS/PVC DWV

56. A watertight pan of corrosion-resistant material is required when a water heater is located in a (an) _____.
    (A) bathroom
    (B) garage
    (C) attic
    (D) basement

57. Where joints are permitted for copper water tubing under a building slab, they shall be brazed and fittings shall be _____.
    (A) cast brass
    (B) wrought copper
    (C) cast bronze
    (D) DWV copper

58. The minimum distance between the center of a water closet and a finished side wall or partition is _____ inches.
    (A) 15
    (B) 30
    (C) 12
    (D) 6

59. When supplied, the finished dam or threshold for a shower, measured from the top of the drain, shall not be less than _____ inch(es).
    (A) 1
    (B) 2
    (C) 3
    (D) 6

60. All shower compartments regardless of shape, excluding accessibility requirements and shower receptors having an overall dimension of not less than 30 inches in width and 60 inches in length, shall be capable of encompassing a minimum _____ inch circle.
    (A) 30
    (B) 36
    (C) 40
    (D) 48

61. The minimum size septic tank for a three-bedroom residence is _____ gallons.
    (A) 750
    (B) 1,000
    (C) 1,200
    (D) 1,500

62. A seepage pit shall have a minimum side wall (not including the arch) of _____.
    (A) 10 feet below inlet
    (B) 20 feet below inlet
    (C) 10 feet below the dome
    (D) 20 feet belowground

63. Flexible water connectors shall be installed in _____.
    (A) accessible locations
    (B) interior locations only
    (C) readily accessible locations
    (D) water softeners only

64. The maximum trap loading for a 2 inch trap is _____ unit(s).
    (A) 1
    (B) 2
    (C) 3
    (D) 4

65. Excluding retrofit installations, the minimum size trap for a shower with a single head is _____ inches.
    (A) 1-1/2
    (B) 2
    (C) 3
    (D) 4

66. Indirect waste pipes less than 15 feet in length shall not be less than the _____ and in no case less than 1/2-inch.
    (A) trap to which it is connected
    (B) largest drain in the building
    (C) size required to drain the fixture in one minute
    (D) drain outlet or tailpiece of the fixture

67. Food preparation sinks shall be indirectly connected to the drainage system by means of a (an) _____.
(A) airgap
(B) floor drain
(C) grease trap
(D) standpipe

68. The minimum size trap for a public (commercial) bar sink is _____ inches.
(A) 1-1/2
(B) 2
(C) 3
(D) 4

69. Seepage pits shall be circular and have an excavated diameter of not less than _____ feet.
(A) 2
(B) 3
(C) 4
(D) 5

70. No plumbing system or portion thereof shall be covered, concealed or put into use until it is inspected and approved by the _____.
(A) Authority Having Jurisdiction
(B) plumbing contractor
(C) general contractor
(D) owner

71. A permit is not required for _____.
(A) replacing water piping
(B) replacing gas piping
(C) capping a sewer
(D) reinstallation of a water closet

72. Waste outlets and fixture tailpieces for kitchen sinks, laundry sinks, and bathtubs shall not be less than _____.
(A) 1-1/2 inches I.D.
(B) 1-1/2 inches
(C) 1-1/4 inches
(D) 1-1/4 inches O.D.

73. The total length of a 1-1/2-inch minimum sized vent pipe that may be installed horizontally is _____ feet.
(A) 20
(B) 30
(C) 40
(D) 60

74. Vent pipes may offset horizontally when the distance above the overflow level of the fixture is not less than _____.
(A) 1 inch
(B) 1 foot
(C) 6 inches
(D) 18 inches

75. Each vent pipe or stack shall extend through its flashings and shall terminate vertically not less than _____ inches above the roof.
(A) 6
(B) 12
(C) 18
(D) 24

76. A factor for determining loss in static pressure from the meter to the highest outlet is _____.
(A) minimum water pressure
(B) elevation above water meter
(C) fixture units
(D) pipe material

77. The minimum size gas pipe supplying a domestic freestanding gas range using natural gas with a pressure drop of 0.5 inches water column, located 60 feet from the meter (1,100 Btu per cubic foot) is _____ inch.
(A) 1/2
(B) 3/4
(C) 1
(D) 3/8

78. Closet bends must be cut off so as to present a _____.
    (A) smooth surface even with the top of the closet ring
    (B) smooth surface even with the floor surface
    (C) surface that can be sealed with putty
    (D) surface that can be painted

79. Leaks in gas piping shall be located by applying _____.
    (A) fire or acid to the exterior of the piping
    (B) noncorrosive leak detection fluid
    (C) ether to the interior of the piping
    (D) smoke to the interior of the piping

80. Caulked joints shall be filled with molten lead to a depth of at least _____ inch(es).
    (A) 1/2
    (B) 3/4
    (C) 1
    (D) 1-1/2

81. The cleanout fitting for a 4-inch diameter drain pipe shall not be smaller than _____ inches.
    (A) 2
    (B) 3
    (C) 3-1/2
    (D) 4

82. A floor drain normally requires a 2-inch trap and a 2-inch waste line. In a combination waste and vent system, the size of the trap and waste line shall be at least _____ inches.
    (A) 2
    (B) 2-1/2
    (C) 3
    (D) 3-1/2

83. Fuel-burning water heaters shall be ensured an adequate supply of _____.
    (A) heat
    (B) air
    (C) base
    (D) A and C

84. Fuel-burning water heaters shall be permitted to be installed in a closet located in the bedroom provided the closet is equipped with a _____.
    (A) deadbolt lock
    (B) 12 inch x 12 inch opening
    (C) metal door
    (D) listed door and self-closing device

85. An attic in which a water heater is installed shall be accessible through an opening of not less than _____.
    (A) 22" x 30"
    (B) 24" x 24"
    (C) 30" x 30"
    (D) 36" x 36"

86. Approved drawn-copper alloy tubing traps may be used on fixtures discharging _____.
    (A) acid waste
    (B) chemical waste
    (C) urinal waste
    (D) domestic sewage

87. Standpipe receptors for clothes washers shall extend above the trap not less than _____ inches.
    (A) 18
    (B) 24
    (C) 30
    (D) 48

88. A horizontal drainage line connecting with other horizontal drainage lines shall enter through _____.
    (A) tapped tee branches
    (B) 45-degree wye branches
    (C) sanitary tee branches
    (D) 60-degree branches

89. Where approved by the Authority Having Jurisdiction, sewer piping 8 inches and larger may slope ______.
    (A) 1/4 inch per foot
    (B) 1/8 inch per foot
    (C) 1/16 inch per foot
    (D) no limit

90. After connecting a building to the public sewer, abandoned private sewage disposal facilities shall be backfilled within ______ days.
    (A) 15
    (B) 30
    (C) 45
    (D) 90

91. PEX tubing shall not be installed within the first ______ inches of piping connected to a water heater.
    (A) 6
    (B) 12
    (C) 18
    (D) 24

92. Pressure-relief valves located inside buildings shall be provided with full-size discharge piping that shall extend ______.
    (A) outside of the building
    (B) under the building
    (C) through an air break to a plumbing fixture
    (D) by direct connection to a fixture tailpiece

93. The valve installed on the cold water supply pipe to each water heater shall be ______.
    (A) a globe valve
    (B) a fullway valve
    (C) a needle valve
    (D) an angle valve

94. The discharge line from a sewage ejector shall be provided with ______.
    (A) a bypass
    (B) a globe valve
    (C) an air relief and siphon breaker
    (D) a backwater valve and gate valve

95. Standard trap seals are required to be no less than/nor more than ______.
    (A) 1 inch/ 3 inches
    (B) 2 inches/ 4 inches
    (C) 3 inches/ 6 inches
    (D) minimum 1 inch/ no maximum

96. Copper tubing and fittings shall be permitted to be used with DWV systems except for ______ wastes.
    (A) condensate
    (B) chemical
    (C) grease
    (D) soil

97. A room has dimensions of 24 feet x 24 feet x 9 feet. Using the Standard Method for calculating combustion air, this space can support a maximum water heater input rating of how many Btu/h?
    (A) 90,000
    (B) 95,000
    (C) 100,000
    (D) 110,000

98. A rainwater catchment system is designed with an underground storage tank. The tank shall be ______.
    (A) not less than 360 gallons capacity
    (B) capable of supporting a hydrostatic load of 60 psf
    (C) not greater than 5,000 gallons capacity
    (D) ballasted or anchored

GENERAL EXAMINATION #3

1. A plumbing permit is required for _____.
   (A) replacing faucet washers
   (B) clearing drain line stoppages
   (C) cleaning hot water tanks
   (D) replacing plumbing fixtures

2. Before any portion of a combination waste and vent system is installed, the code requires that _____.
   (A) the owner give his consent in writing
   (B) construction documents be approved by the Authority Having Jurisdiction
   (C) trap primers be installed for each trap
   (D) a separate connection be made to the public sewer

3. The maximum number of water closets on a 3 inch horizontal pipe or stack shall not exceed _____.
   (A) three
   (B) four
   (C) five
   (D) six

4. Sleeves shall be provided to protect piping through _____ and _____ walls, and concrete floors.
   (A) floors, concrete
   (B) wood, concrete
   (C) concrete, masonry
   (D) none of the above

5. The enlargement of a 3 inch closet bend or stub to 4 inches is _____.
   (A) not considered an obstruction
   (B) not an accepted engineering practice
   (C) allowed when necessary due to structural conditions
   (D) considered an obstruction and should not be used

6. All trenches deeper than the footing of any building or structure and paralleling the same shall be located not less than _____ from the bottom exterior edge of the footing or as approved.
   (A) 30 degrees
   (B) 45 degrees
   (C) 5 feet
   (D) 8 feet

7. Four inch cast-iron soil stacks shall be supported at _____.
   (A) 10 foot intervals
   (B) 5 foot intervals
   (C) the base and each floor not to exceed 15 feet
   (D) each hub and fitting

8. The water test for drainage and venting systems shall be not less than _____.
   (A) a 10 foot head of water
   (B) a 5 foot head of water
   (C) 10 psi
   (D) 25 psi

9. The total drainage fixture unit value for a single-family home with one kitchen sink, one laundry sink, one bathtub, one lavatory, and one 1.6 gravity tank-type water closet is _____ F.U.
   (A) 15
   (B) 13
   (C) 11
   (D) 10

10. ABS-DWV piping shall be limited to ______.
    (A) structures not exceeding three stories above grade
    (B) non combustible commercial occupancies only
    (C) dwellings and apartment hotels
    (D) no limitations if installed in accordance with applicable standards in Table 701.2 and Chapter 14

11. Copper tube for drainage and venting systems shall be at least ______.
    (A) Type DWV
    (B) Type K
    (C) Type L
    (D) Type M

12. An acceptable fitting for connecting a vertical drainage line to a horizontal drain line is a ______.
    (A) sanitary tee
    (B) tapped tee
    (C) low heel 1/4 bend
    (D) wye and 1/8 bend

13. With some exceptions, each horizontal drainage pipe shall be provided with a cleanout at its ______.
    (A) upper terminal
    (B) lower terminal
    (C) wye connection
    (D) midpoint

14. The cleanout opening for a 4-inch house drain shall not be smaller than ______ inches.
    (A) 2-1/2
    (B) 3-1/2
    (C) 3
    (D) 4

15. In general, horizontal drainage piping shall be run at a uniform slope of not less than which of the following percent?
    (A) 1
    (B) 2
    (C) 3
    (D) 4

16. The minimum size discharge pipe from an ejector in a single-dwelling unit having a water closet connected thereto shall be not less than ______ inches.
    (A) 1-1/2
    (B) 2
    (C) 3
    (D) 4

17. The maximum number of kitchen sinks allowed on a 2-inch vertical waste pipe is ______.
    (A) four
    (B) six
    (C) eight
    (D) ten

18. The maximum length for a 1-1/2-inch vertical vent pipe is ______.
    (A) 45
    (B) 60
    (C) 120
    (D) 180

19. Vent terminals in locations having minimum design temperatures below 0°F shall be not less than ______ inches.
    (A) 3
    (B) 4
    (C) 1-1/2
    (D) 2

20. Vent piping installed horizontally and a minimum of 6 inches above the the overflow rim of the fixture shall be ______.
    (A) level or so graded as to drip back to the drain it serves
    (B) offset
    (C) increased one pipe size
    (D) brass or copper

21. Except for those serving refrigeration coils and ice machines, an indirect waste shall not exceed _____ feet.
(A) 15 (C) 50
(B) 25 (D) 75

22. The maximum temperature of water discharging under pressure directly into the drainage system is _____.
(A) 140°F (C) 100°C
(B) 180°F (D) 212°F

23. The trap serving a floor sink with a 2-inch tailpiece as part of a combination waste and vent system shall not be less than _____ inches.
(A) 2 (C) 3
(B) 2-1/2 (D) 3-1/2

24. One trap may serve a set of three single-compartment sinks where the waste outlets are not more than _____ inches apart and the trap is centrally located.
(A) 24 (C) 18
(B) 30 (D) 36

25. The vertical distance between a fixture outlet and the trap weir shall not exceed _____ inches.
(A) 18 (C) 30
(B) 24 (D) 36

26. The developed length of a 2-inch trap arm shall be a maximum of _____.
(A) 2 feet, 6 inches (C) 5 feet
(B) 3 feet, 6 inches (D) 6 feet

27. Fixture traps shall have a liquid seal of not less than _____ inch(es).
(A) 2 (C) 6
(B) 4 (D) 1

28. Closet bends or stubs shall be cut off so as to present a smooth surface and even with the _____.
(A) top of the floor (C) flow line of the pipe
(B) top of the closet ring (D) floor tile or linoleum

29. A fixture tailpiece for a kitchen sink shall be not less than _____ inches in diameter.
(A) 1-1/2 (C) 1-1/4
(B) 1-3/8 (D) 1-3/4

30. When using Appendix A sizing, the minimum residual water pressure that shall be supplied at the highest fixtures in a supply system containing flush tank shall not be less than _____ psi or the required minimum residual pressure of a fixture if higher.
(A) 0.434 (C) 4.34
(B) 30 (D) 15

31. Elastomeric gasketed and rubber gasket joints shall comply with _____.
(A) manufacturer's approval (C) the Authority Having Jurisdiction
(B) referenced standards Table 1701.1 (D) special joints

32. Cast-iron soil pipe shall not be joined by _____.
(A) brazing (C) compression-type joints
(B) caulking (D) shielded couplings

33. Approved unions may be used at any point in the _____.
   (A) drainage system
   (B) vent system
   (C) gas system
   (D) water supply system

34. A bathtub drain assembly with slip joint connections that will be concealed shall be provided with _____.
   (A) a 22 B&S gauge waste and overflow
   (B) lead slip joint washers in lieu of rubber
   (C) a minimum 12-inch x 12-inch access panel
   (D) a 1-3/8 inch O.D. waste and overflow

35. The distance from the center of any water closet to a side wall or partition shall not be less than _____ inches.
   (A) 12
   (B) 15
   (C) 24
   (D) 30

36. No shower stall or receptor shall have a finished interior unable of encompassing a 30-inch circle, except where _____.
   (A) showers designed to be accessible
   (B) showers having an overall dimension of not less than 30 inches in width and 60 inches in length
   (C) existing bathtub is replaced by a shower
   (D) both A and B apply

37. The vacuum breaker, which is a part of a flushometer valve serving a water closet, shall be installed with the critical level not less than _____ inch(es) above the overflow rim of the bowl or the distance according to its listing.
   (A) 1
   (B) 6
   (C) 12
   (D) 18

38. Valves on the discharge side of each water meter shall be _____.
   (A) fullway valves
   (B) fullway gate valves only
   (C) globe valves or fullway gate valves
   (D) globe valves or butterfly valves

39. Water pressure regulators are required where the static water pressure is exceeding _____ psi.
   (A) 80
   (B) 100
   (C) 125
   (D) 150

40. The building supply size shall be determined by various criteria, but in no case shall it be less than _____ inch.
   (A) 1/2
   (B) 3/4
   (C) 1
   (D) there are no restrictions

41. The minimum size of any building sewer shall be determined on the basis of the _____.
   (A) number of bedrooms
   (B) number of bathrooms
   (C) type of building
   (D) total fixture units

42. In a straight run of any building sewer, cleanouts are required at intervals of not to exceed _____.
   (A) 50 feet
   (B) 100 feet
   (C) five full lengths of pipe
   (D) 10 full lengths of pipe

43. Every abandoned building sewer shall be plugged or capped within _____ feet of the property line.
   (A) 2
   (B) 5
   (C) 6
   (D) 10

44. The minimum cover for underground gas piping systems installed where external damage to the pipe is not likely to result shall be _____ inches.
(A) 12
(B) 18
(C) 24
(D) 30

45. The size of the supply piping outlet for any gas appliance shall not be less than _____.
(A) the inlet connection of that appliance
(B) 1/2 inch
(C) 3/4 inch
(D) 3/8 inch

46. Indoor gas hose connectors used to connect laboratory, shop and ironing appliances or equipment requiring mobility shall be of minimum length and shall not exceed _____ feet.
(A) 2
(B) 3
(C) 6
(D) 5

47. Each combustion air opening used to connect indoor spaces shall have a free area not less than _____ square inches.
(A) 50
(B) 80
(C) 100
(D) 160

48. Unless listed as flammable vapor ignition resistant, gas water heaters in garages shall be installed so that burners and burner-ignition devices are located not less than _____ inches above the floor.
(A) 18
(B) 24
(C) 12
(D) 30

49. The total fall of a 4-inch house sewer 150 feet long and installed at a grade of 1/8 inch per foot is most nearly _____.
(A) 15.5
(B) 18.75
(C) 30.25
(D) 37

50. Expansion and contraction of drain and vent pipes may be provided for by use of _____.
(A) spring hangers
(B) expansion joints
(C) a plumber's loop
(D) a return bend

51. The inspection of a gas piping system prior to covering or concealing such piping is called a _____.
(A) hydro piping inspection
(B) rough piping inspection
(C) routine inspection
(D) site inspection

52. The static water pressure at the water meter is 80 psi. At a supply outlet 40 feet above the meter, the static water pressure would be approximately _____ psi.
(A) 40
(B) 60
(C) 80
(D) 120

53. The maximum number of water closets allowed on a 3 inch horizontal branch is _____.
(A) three
(B) four
(C) five
(D) six

54. The discharge line from a sewage ejector may be provided with a _____.
(A) backwater valve and a check valve
(B) backwater valve and a gate valve
(C) swing check valve and a ball valve
(D) both B and C apply

55. In a special venting installation for island sinks, the return bend used under the drain board may be a _____.
    (A) 45-degree, 90-degree, 45-degree elbow
    (B) 30-degree, 60-degree, 90-degree elbow
    (C) swivel 90-degree elbow
    (D) two 90-degree elbows

56. A food waste disposer installed in a set of restaurant sinks shall be connected to a _____.
    (A) single trap
    (B) separate trap
    (C) diverter tee
    (D) grease trap

57. Each plumbing fixture trap shall be protected against siphonage and back pressure by means of a _____.
    (A) vent pipe
    (B) deep seal trap
    (C) check valve
    (D) drum trap

58. A joint made by firmly packing with oakum and filled with molten lead to a depth of not less than one inch, is used with _____.
    (A) steel pipe
    (B) plastic pipe
    (C) cast-iron hub and spigot soil pipe
    (D) cast-iron hubless pipe

59. Drains for gutters in public or gang shower rooms shall be spaced not more than _____ feet apart.
    (A) 10
    (B) 8
    (C) 12
    (D) 16

60. Potable water supply to beverage dispensers shall be protected by either an airgap or _____.
    (A) an airbreak
    (B) a reduced-pressure backflow prevention device
    (C) a stainless steel reduced-pressure backflow prevention device
    (D) a vented backflow preventer complying with ASSE 1022

61. When using Table 610.4 to size a water supply system serving flushometer valves, the first water closet is assigned a value of _____ fixture units.
    (A) 40
    (B) 30
    (C) 20
    (D) 15

62. Except as otherwise approved by the Authority Having Jurisdiction, velocities in water piping systems designed using Appendix A shall not exceed _____ feet per second.
    (A) 10
    (B) 15
    (C) 20
    (D) 30

63. The minimum septic tank capacity, in gallons, for a single-family dwelling with two bedrooms is _____ gallons.
    (A) 500
    (B) 750
    (C) 1,000
    (D) 1,200

64. Joints at sections of single wall vent connector piping shall be fastened with _____.
    (A) duct tape
    (B) sheet metal screws
    (C) plumbers tape
    (D) 26 gauge wire

65. Gas piping inside any building shall not be installed in or through a (an) _____.
    (A) elevator shaft
    (B) clothes chute
    (C) gas vent
    (D) all of the above

66. The finished floor of a shower receptor shall slope uniformly from the sides toward the drain with not less than _____ inch per foot not more than _____ inch per foot.
(A) 1/8, 1/4
(B) 1/8, 1/2
(C) 1/2, 3/8
(D) 1/2, 1

67. A sewer line installed horizontally will require additional cleanouts for each aggregate change of direction _____.
(A) less than 135 degrees
(B) exceeding 135 degrees
(C) less than 90 degrees
(D) exceeding 90 degrees

68. One requirement for backwater valves is that they shall _____.
(A) be installed in the house drain
(B) be installed belowground
(C) screen solids and pass liquid wastes
(D) remain open during periods of low flows

69. A _____ is a prohibited fitting and shall not be used as a drainage fitting.
(A) side inlet quarter bend
(B) double sanitary tee
(C) double wye
(D) 4 inch x 3 inch closet bend

70. Hubless cast-iron piping in 5 foot lengths, installed horizontally aboveground, shall be supported at _____.
(A) each joint
(B) 5 foot intervals
(C) every other joint
(D) 10 foot intervals

71. The Authority Having Jurisdiction shall be permitted to require that a request for inspection be filed not less than _____ hours before such inspection is desired.
(A) 8
(B) 12
(C) 24
(D) 36

72. Building sewers shall be tested by filling with water _____.
(A) to at least a 10 foot head
(B) from the lowest to the highest point
(C) or 5 pounds per square inch air
(D) and retain same for 15 minutes

73. Drainage fittings shall not be made of _____
(A) Stainless Steel 316L
(B) brass
(C) ABC or PVC
(D) lead

74. In a private installation, two 1.6 tank-type water closets, two lavatories, one kitchen sink, one clothes washer, one bathtub, and one shower (private installation) have a total waste fixture unit count of _____ fixture units.
(A) 17
(B) 20
(C) 24
(D) 30

75. The common waste pipe serving four clothes washers in a battery will have a total of _____ fixture units.
(A) 8
(B) 12
(C) 24
(D) 16

76. Which of the following is a false statement with regard to combination waste and vent systems?
(A) such systems are self-scouring
(B) trap two pipe sizes larger than fixture tailpiece
(C) vent pipe one-half the inside cross-sectional area of drain pipe served
(D) any branch more than 15 feet in length is separately vented

77. A 6 inch type B gas vent, with a listed cap terminating 1 foot above a flat roof shall be not less than _____ feet horizontally from a vertical wall or similar obstruction.
    (A) 4
    (B) 8
    (C) 10
    (D) 12

78. Single-wall metal pipe for vent connectors shall be constructed of galvanized sheet steel not less than _____ of an inch thick.
    (A) 0.304
    (B) 0.0304
    (C) 0.403
    (D) 0.413

79. To obtain the cubic feet per hour of gas required, divide the Btu input of appliance by _____.
    (A) 1,000
    (B) 1,100
    (C) the average Btu heating value per cubic foot of the gas in the area of the installation
    (D) none of the above

80. Detection of leaks in gas piping systems shall be located by means of _____.
    (A) an open flame
    (B) wintergreen oil
    (C) a noncorrosive leak detection fluid
    (D) a water pressure test

81. The inspection of gas piping that includes a determination that the gas piping size, material, and installation meet the requirement of the code is called a _____.
    (A) rough piping inspection
    (B) final piping inspection
    (C) topout inspection
    (D) gas company inspection

82. Relating to cesspool construction, where a strata of gravel, or equally pervious material, of 4 feet in thickness is found, the depth of a cesspool sidewall shall not exceed _____ feet below the inlet.
    (A) 5
    (B) 8
    (C) 10
    (D) 20

83. The maximum rainfall rate for San Francisco, California, in the last 100 years is _____ inches per hour.
    (A) 4.5
    (B) 3.7
    (C) 2.0
    (D) 1.5

84. A pressure-type vacuum breaker is a device designed to be used for _____.
    (A) controlling water pressure
    (B) developing a vacuum
    (C) preventing back-siphonage
    (D) eliminating water hammering

85. Wall-mounted water closets shall be rigidly supported by _____.
    (A) wall brackets
    (B) metal supporting members (carriers)
    (C) closet bends
    (D) floor flanges

86. Traps for bathtubs shall be constructed of which approved material?
    (A) die-cast zinc
    (B) aluminum
    (C) steel
    (D) brass

87. Unless other means acceptable to the Authority Having Jurisdiction are used, appliances located on the roof require an inside means of access to the roof for buildings exceeding _____ feet in height.
    (A) 15
    (B) 18
    (C) 20
    (D) 25

88. Except for urinals, drawn-copper alloy tubing traps not less than 17 B&S gauge may be installed on fixtures discharging domestic sewage provided that they are ______.
(A) installed 1 foot above grade
(B) not used in condominiums
(C) used only in single-family residences
(D) exposed and readily accessible

89. The developed length between a closet flange serving a water closet and its vent shall not exceed ______ feet.
(A) 6
(B) 4
(C) 10
(D) 15

90. A ______ shall be installed for special conditions.
(A) full "S" trap
(B) drum trap
(C) bell trap
(D) crown vented trap

91. Indirect waste pipes exceeding 5 feet in length shall be ______.
(A) trapped
(B) vented
(C) painted
(D) threaded

92. The limits of a clothes washer standpipe height above its trap is ______ inches.
(A) 18 to 30
(B) 12 to 18
(C) 18 to 36
(D) 24 to 30

93. Each drainage stack that extends 10 or more stories shall be served by ______.
(A) an overflow drain stack
(B) a velocity indicator
(C) a parallel vent stack
(D) an island vent

94. Indirect waste piping serving ice machines need not be larger in diameter than the drain outlet or tailpiece of the fixture, but in no case less than ______ inch(es).
(A) 1/2
(B) 3/4
(C) 1
(D) 2

95. A bathroom group shall be permitted to be made up of ______.
(A) water closet, tub/shower and lavatory
(B) bidet, lavatory and shower
(C) shower, bidet and emergency floor drain
(D) all of the above meet the definition

96. Air conditioning condensate waste shall not connect indirectly through an air gap or air break to a ______.
(A) floor drain
(B) bathtub overflow
(C) plumbing fixture tailpiece
(D) leach pit

97. Under what conditions are multiple buildings allowed to 'share' aggregate cross-sectional venting areas to satisfy the requirements of the total vent pipe areas being not less than the area of the largest required building sewer?
(A) one ownership
(B) common building sewer
(C) buildings occupy the same lot
(D) A, B & C are all required

PLUMBING MATHEMATICS EXAMINATION #1

Mathematics is an important part of the plumbing trade. The ability to accurately measure sections of pipe, equipment, and supports is essential. This includes simple hydraulic calculations involving volume, velocity, and weight of water. The questions in this examination are limited to mathematics that are basic to the trade.

Formulas used to calculate correct answers are found in the Useful Tables following the Plumbing Mathematics Examinations. The important thing to remember is to use the formula correctly. Particular attention should be given to the units involved.

Avoid mixing inches with feet, square feet, or square inches. Try to do all calculations in a clear and systematic order so that if your answer is wrong you can recheck your calculations for errors. An excellent method in understanding the problem is to make a sketch of what is given. Make sure you have all units in the proper order and are using the correct calculation method or formula.

1. Atmospheric pressure (barometric pressure) at sea level is most nearly _____.
   (A) 8.34 lbs.
   (B) 14.7 psi
   (C) 0.434 psi
   (D) 62.4 psi

2. One cubic foot contains _____ cubic inches.
   (A) 144
   (B) 231
   (C) 1,728
   (D) 61.2

3. The volume of water in one cubic foot is approximately _____.
   (A) 7-1/2 gal.
   (B) 6-1/2 gal.
   (C) 28 qt.
   (D) 10 qt.

4. The pressure at the base of a column of water 1-inch high is approximately _____.
   (A) 0.434 psi
   (B) 4.335 psi
   (C) 0.036 psi
   (D) 2.31 psi

5. One gallon of water weighs approximately _____ pounds.
   (A) 8.34
   (B) 4.34
   (C) 2.31
   (D) 4.91

6. The pressure required to raise a column of water 200 feet would be (neglect friction losses) approximately _____ psi.
   (A) 100
   (B) 200
   (C) 86
   (D) 43

7. The full length of a house sewer is 100 feet and the total fall is 25 inches. The slope of the sewer is approximately _____ inch/foot.
   (A) 1/4
   (B) 1/2
   (C) 1/8
   (D) 3/8

8. One cubic foot of water weighs approximately _____ pounds.
   (A) 144
   (B) 231
   (C) 62.4
   (D) 43.4

9. A temperature of 212° Fahrenheit is equivalent to _____ ° Celsius.
   (A) 180
   (B) 100
   (C) 273
   (D) 212

10. One square foot contains _____ square inches.
   (A) 144
   (B) 100
   (C) 36
   (D) 48

11. Doubling the diameter of a pipe will _____.
   (A) double the area
   (B) increase friction losses
   (C) increase the area four times
   (D) not increase volume flow

12. The scale on a set of plans states that 1/8 inch = 1 foot. The actual length of a waterline with plan dimensions of 5 inches would be _____ feet.
   (A) 8
   (B) 5
   (C) 64
   (D) 40

13. The cross-sectional area of a 2 inch diameter pipe is approximately _____ square inches.
   (A) 3.1416
   (B) 6.2832
   (C) 4.9087
   (D) 12.567

14. A cold water pipe is to be offset 4 feet using 45-degree elbows. The length of the diagonal pipe will be approximately _____.
   (A) 5 ft.
   (B) 8 ft.
   (C) 5 ft., 7 in.
   (D) 6 ft., 6 in.

15. A 4-inch diameter pipe 100 feet in height is filled to the top with water. The weight at the base of the stack would be approximately _____ pounds.
   (A) 100
   (B) 400
   (C) 545
   (D) 43.4

16. The area of a circle can be computed by using the formula _____.
   (A) S=0.7854 D
   (B) D=1.2732 S
   (C) n=3.1416
   (D) $A=\pi d^2/4$

17. The circumference and the area are numerically the same for a circle with a diameter of _____ inches.
   (A) 2
   (B) 3
   (C) 4
   (D) 5

18. The area of a square is 196 square feet. One side has a length of _____.
   (A) 83 in.
   (B) 49 in.
   (C) 14 ft.
   (D) 49 ft.

19. A rectangular tank, including top, 60ft. x 20ft. x 10ft., is to be covered inside with a sheet copper liner. The total surface to be covered is _____ square feet.
   (A) 2,800
   (B) 3,600
   (C) 3,200
   (D) 4,000

20. A cesspool is 48 inches in diameter and 20 feet deep. If filled to the top with water, the volume in gallons would be approximately _____ gallons.
    (A) 1,880
    (B) 1,808
    (C) 2,000
    (D) 2,500

21. The area of four 2-inch diameter pipes is equivalent to the area of _____ inch diameter pipe(s).
    (A) one 3
    (B) one 4
    (C) one 8
    (D) eight 1

22. The sum of the following pipe measurements 8'-6" + 6'-7" + 10'-4" + 9'-2" + 4'-4" is _____.
    (A) 37'
    (B) 38'
    (C) 36'-23"
    (D) 38'-11"

23. The sides of a right triangle are 3 feet and 4 feet. The hypotenuse would be approximately _____ feet.
    (A) 5
    (B) 7
    (C) 12
    (D) 25

(Image included to help understand strap use for question 24.)

24. A fuel tank 5 feet in diameter and 60 feet long is to be supported from 4 a steel beams with 4-U straps. The total length of strap material will be at least _____ feet.
    (A) 13
    (B) 26
    (C) 39
    (D) 52

25. Of the following, the one with the highest rate of flow is _____.
    (A) 30,000 gal. per 10 hr.
    (B) 1,000 gal. per hr.
    (C) 20 gal. per min.
    (D) 1 gal. per sec.

26. A tank has a volume of 864 cubic feet. If one cubic foot equals 7.48 gallons, the tank will hold approximately _____ gallons.
    (A) 11.55
    (B) 6462.72
    (C) 64.62
    (D) 646.27

27. A pump tank has a diameter of 40 inches. To keep the pump from short-cycling, the floats must be placed to allow a minimum of 50 gallons of water discharge. What is the minimum distance between the start and stop floats in the tank?
    (A) 9.2 inches
    (B) 19.2 inches
    (C) 29.2 inches
    (D) 39.2 inches

28. A natural gas pressure of 8 inches water column is most nearly _____ psi.
    (A) .036
    (B) .434
    (C) .289
    (D) .504

29. Where a 1-inch diameter pipe is replaced with a 2-inch diameter pipe, the cross-sectional area will _____.
    (A) increase two times
    (B) increase four times
    (C) remain the same
    (D) decrease slightly

30. The aggregate cross-sectional area of all vents for a system with a required 4 inch sewer shall be at least _____ square inches.
    (A) 7.06
    (B) 12.57
    (C) 18
    (D) 4

PLUMBING MATHEMATICS EXAMINATION #2

For the most accurate results, you will need a calculator with the basic mathematical functions (including square, pi, percent, and square root). Do not round numbers until the final answer. Round the final answer to two decimal places where appropriate.

Include the proper units (square feet, cubic feet, inches, feet, pounds, gallons, etc., are some examples of units). Be careful not to mix units.

1. What is the surface area of a rectangular tank having a length of 25 feet, a width of 15 feet, and a height of 8 feet? _____
2. What is the volume of the tank in question #1? _____
3. A rectangular tank 4'-6" wide, 9'-3" long, and 22'-9" high is to be covered inside with a sheet copper liner. The surface to be covered is how many square feet? _____
4. What is the volume of the tank in question #3? _____
5. What is the volume of a cylinder that has a diameter of 3'-6" and a height of 15'-3"? _____
6. What is the weight of water in the cylinder in question #5? _____
7. What is the area of a triangle with a base of 8" and a height of 15"? _____
8. What is the area of a 45-45-90 degree triangle having two equal sides of 15 inches? _____
9. What is the length of the hypotenuse of the triangle in question #8? _____
10. What is the length of the hypotenuse of a 90° triangle with sides of 5" and 10"? _____
11. What is the cross-sectional area of a 20 ft. diameter pipe? _____
12. What is the length of a side of a square with an area of 324 sq. in.? _____
13. A cylinder has a diameter of 48 inches and a height of 20 feet. What is the area of a liner needed to cover the interior surface of the cylinder? _____
14. What is the area of a triangle with a base of 20" and a height of 10"? _____
15. What is the cross-sectional area of a pipe with a diameter of 15'? _____
16. What is the length of the hypotenuse of a 45-45-90 degree triangle having two equal sides of 12 inches? _____
17. What is the area of the triangle in question #16? _____
18. What is the length of the hypotenuse of a 90° triangle with sides of 8" and 16"? _____

19. What is the length of the side of a square with an area of 361 sq. in.? _____

20. A cylinder has a diameter of 36 inches and a height of 11 feet. What is the area of a liner needed to cover the interior surface of the cylinder? _____

21. What is the area of a liner (including all surfaces) for a rectangular tank having a length of 7 feet, a width of 18 feet, and a height of 30 feet? _____

(Image included to help understand strap use for questions 22 & 23.)

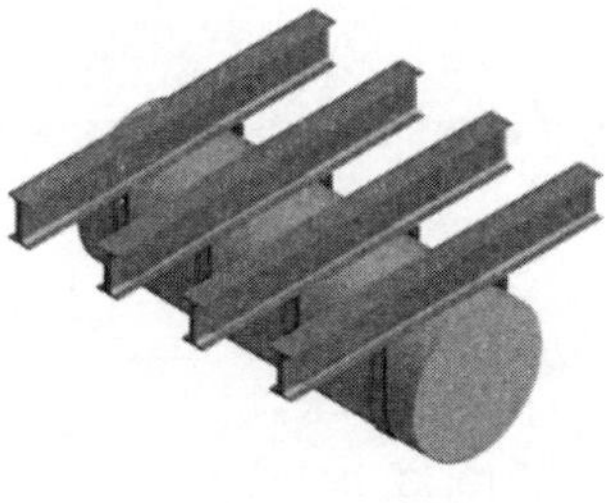

22. A cylinder with a diameter of 36" and length of 20' is to be hung from three I-beams with three U-straps. What is the minimum length of strapping around the cylinder? _____

23. A cylinder with a diameter of 90" and length of 30' is to be hung from five I-beams with five U-straps. What is the minimum length of strapping around the cylinder? _____

24. A pipe is to be offset on a 45° angle between two parallel lines 87" apart. The length of pipe needed to complete this offset would be most nearly? _____

USEFUL TABLES

MATH FORMULAS

LEGEND		
A = area	F = Fahrenheit	r = radius
b = base	H = Hypotenuse	s = side
C = Celsius	h = height (altitude)	V = volume
d = diameter	l = length	w = width

1"	.083 ft.
π Constant	$\pi = 3.14159$
Area of circle	$A = \pi r^2$
Area of a parallelogram	$A = bh$
Area of a rectangle	$A = lw$
Area of a square	$A = s^2$
Area of trapezoid	$A = 1/2(b_1+b_2)h$
Area of a triangle	$A = 1/2(bh)$ or $a=\frac{bh}{2}$
Average atmospheric air pressure at sea level (59°F)	29.9212 inches hg or 14.696 psi
Circumference of a circle	$C = \pi d$
Cubic yard	27 cubic feet
Hypotenuse of an isosceles right triangle (45-45-90 triangle)	$H = s \times \sqrt{2}$
Length of strap to support a tank	$l = 1/2(2\pi r) + 2r$
Perimeter of a rectangle.	$P = 2(l+w)$
Pythagorean Theorem (to find hypotenuse)	$a^2 + b^2 = c^2$
Surface area of a cylinder	$A = 2\pi r^2 + 2\pi rh$
Surface area of a rectangular box	$A = 2(hw + hl + wl)$
Temperature conversion °C to °F	$°F = (9/5°C)+32$
Temperature conversion °F to °C	$°C = 5/9(°F-32)$
Volume of cylinder	$V = (\pi r^2)h$
Volume of prism	$V = b_{(area)}h$
Volume of pyramid	$V = 1/3b_{(area)}h$
Volume of rectangular prism	$V = lwh$
Volume of sphere	$V = 4/3\pi r^3$

PROPERTIES OF WATER

1 cubic foot weighs	62.427 lbs.
1 gallon weighs	8.3453 lbs.
1 cubic foot contains	7.4805 gallons
1 gallon contains	231 cubic inches
1 cubic foot contains	1728 cubic inches
Water pressure increases / decreases	0.434 psi / foot elevation
Water pressure increases / decreases	0.0361 psi / inch elevation

CONVERSION TABLES

MULTIPLY	BY	TO OBTAIN
Acres	43 560	Square feet
Acre-feet	43 560	Cubic feet
Acre-feet	325 851	Gallons (U.S. liquid)
Atmosphere (standard) (atm)	76.0	Centimeters of mercury (0°C)
Atmosphere (standard)	33.90	Feet of water (4°C)
Atmosphere (standard)	29.92	Inches of mercury
Atmosphere (standard)	101.32501	KiloPascals (kPa)
Atmosphere (standard)	14.70	Pounds-force/square inch
British thermal units (Btu)	1055.055	Joules (J)
Btus/hour	0.000293	Kilowatts (kW)
Btus/hour	0.293	Watts (W)
Btus/minute	12.97	Foot pounds-force/second
Btus/minute	0.02358	Horsepower (hp) (international)
Centimeters (cm)	0.3937	Inches
Centimeters of mercury (0°C)	0.01316	Atmosphere (standard)
Centimeters of mercury (0°C)	0.446	Feet of water (4°C)
Centimeters of mercury (0°C)	27.84	Pounds-force/square feet
Centimeters of mercury (0°C)	0.1934	Pounds-force/square inch
Cubic feet (ft^3)	1728	Cubic inches
Cubic feet	0.0283	Cubic meters (m^3)
Cubic feet	0.03704	Cubic yards
Cubic feet	7.48052	Gallons (U.S. liquid)
Cubic feet	29.92	Quarts (U.S. liquid)
Cubic feet/minute (ft^3/min)	0.000472	Cubic meters/second
Cubic feet/minute	0.1247	Gallons/second
Cubic feet/minute	0.47194	Liters/second (L/s)
Cubic feet/second (ft^3/s)	646 316.89	Gallons/day
Cubic feet/second	448.831	Gallons/minute
Cubic yards (yd^3)	27	Cubic feet
Cubic yards	201.97	Gallons (U.S. liquid)
Degrees	0.0174	Rads
Feet (ft)	304.8	Millimeters
Feet of water (4°C)	0.0295	Atmosphere (standard)
Feet of water (4°C)	0.8827	Inches of mercury (0°C)
Feet of water (4°C)	62.43	Pounds-force/square feet
Feet of water (4°C)	0.4335	Pounds-force/square inch
Feet/minute (ft/min)	0.01667	Feet/second
Feet/minute	0.01136	Miles/hour
Feet/second (ft/s)	0.3048	Meters/second (m/s)
Feet/second	0.6818	Miles/hour
Feet/second	0.01136	Miles/minute
Foot pounds-force (ft•lbf)	1.355	Joules
Gallons (U.S. liquid)	231	Cubic inches
Gallons (U.S. liquid) (gal)	0.003785	Cubic meters
Gallons (U.S. liquid)	0.1337	Cubic feet

MULTIPLY	BY	TO OBTAIN
Gallons (U.S. liquid)	3.785	Liters
Gallons (U.S. liquid)	4	Quarts (U.S. liquid)
Gallons/minute (gpm)	8.0208	Cubic feet/hour
Gallons/minute	0.00223	Cubic feet/second
Gallons/minute	0.06309	Liters/second
Grains (gr)	0.00006479	Kilograms (kg)
Inches (in)	2.54	Centimeters
Inches of mercury (0°C)	0.03342	Atmosphere (standard)
Inches of mercury (0°C)	1.133	Feet of water (4°C)
Inches mercury (0°C)	3.3863	KiloPascals (kPa)
Inches of mercury (0°C)	0.4912	Pounds-force/square inch
Inches of water (4°C)	0.002458	Atmosphere (standard)
Inches of water (4°C)	0.07356	Inches of mercury (0°C)
Inches of water (4°C)	5.202	Pounds-force/square feet
Inches of water (4°C)	0.03613	Pounds-force/square inch
KiloPascals (kPa)	0.145038	Pounds-force/square inch
Liters	61.02	Cubic inches
Liters (L)	0.001	Cubic meters
Liters	0.2642	Gallons (U.S. liquid)
Miles	5280	Feet
Miles/hour (mi/h)	88	Feet/minute
Miles/hour	1.467	Feet/second
Millimeters (mm)	0.1	Centimeters
Millimeters	0.03937	Inches
Millimeter	0.001	Meters
Ounces (oz)	0.02834	Kilograms
Pounds (lb)	0.45359	Kilograms
Pounds/cubic foot (lb/ft^3)	16.0184	Kilograms/cubic meter (kg/m^3)
Pounds/square inch (lb/in^2)	703.1	Kilograms-force/square meter (kg/m^2)
Pounds/square foot (lb/ft^2)	4.882427	Kilograms-force/square meter (kg/m^2)
Pounds-force (lbf)	4.4482	Newtons (N)
Pounds-force/square inch (psi)	0.06805	Atmosphere (standard)
Pounds-force/square inch	2.307	Feet of water (4°C)
Pounds-force/square inch	2.036	Inches of mercury (0°C)
Pounds-force/square inch	6.89476	KiloPascals
Quarts (U.S. dry) (dry qt)	67.20	Cubic inches
Quarts (U.S. liquid) (liq qt)	57.75	Cubic inches
Square feet (ft^2)	144	Square inches
Square feet	0.0929	Square meters
Square inches (in^2)	0.000645	Square meters
Square miles (mi^2)	640	Acres
Square yards (yd^2)	9	Square feet
Temperature (°C) + 17.28	1.8	Temperature (°F)
Temperature (°F) – 32	$\frac{5}{9}$	Temperature (°C)
Tons (short)	2000	Pounds
Water column (1 inch)	0.24908	KiloPascals

AREAS AND CIRCUMFERENCES OF CIRCLES

DIAMETER		CIRCUMFERENCE		AREA	
Inches	mm	Inches	mm	Inches^2	mm^2
⅛	6	0.40	10	0.01227	8.0
¼	8	0.79	20	0.04909	31.7
⅜	10	1.18	30	0.11045	71.3
½	15	1.57	40	0.19635	126.7
¾	20	2.36	60	0.44179	285.0
1	25	3.14	80	0.7854	506.7
1¼	32	3.93	100	1.2272	791.7
1½	40	4.71	120	1.7671	1140.1
2	50	6.28	160	3.1416	2026.8
2½	65	7.85	200	4.9087	3166.9
3	80	9.43	240	7.0686	4560.4
4	100	12.55	320	12.566	8107.1
5	125	15.71	400	19.635	12 667.7
6	150	18.85	480	28.274	18 241.3
7	175	21.99	560	38.485	24 828.9
8	200	25.13	640	50.265	32 428.9
9	225	28.27	720	63.617	41 043.1
10	250	31.42	800	78.540	50 670.9

EQUAL PERIPHERIES

$S = 0.7854\ D$

$D = 1.2732\ S$

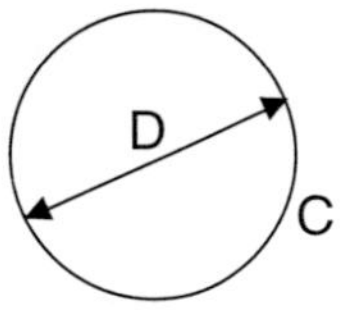

$S = 0.8862\ D$

$D = 1.1284\ S$

$S = 0.2821\ C$

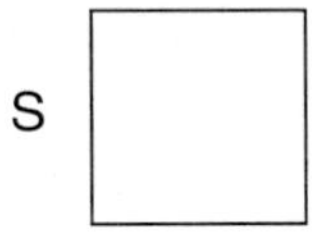

EQUAL AREAS

Area of square (S') = 1.2732 x area of circle

Area of square (S) = 0.6366 x area of circle

$C = pD = 2pR$

$C = 3.5446 \sqrt{area}$

$D = 0.3183\ C = 2R$

$D = 1.1283 \sqrt{area}$

$Area = pR^2 = 0.7854\ D^2$

$Area = 0.07958\ C^2 = \frac{pD^2}{4}$

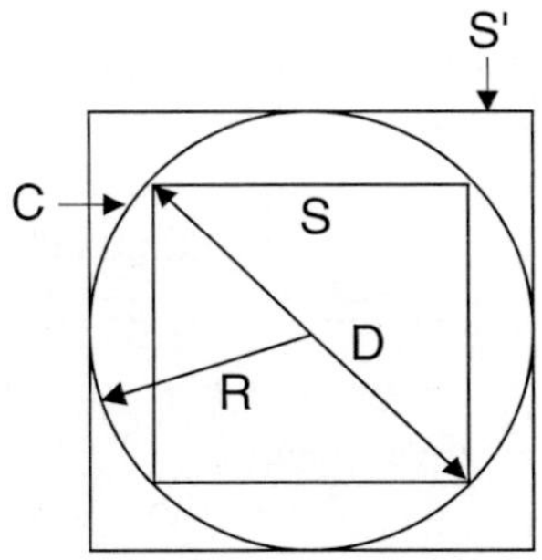

$p = 3.1416$

FLOW IN PARTLY FILLED (ONE-HALF FULL) PIPES
(BASED ON MANNING'S FORMULA WITH n = .012)

gpm x 6.31 = L/s
ft/sec x .305 = m/s
in. x 25.4 = mm

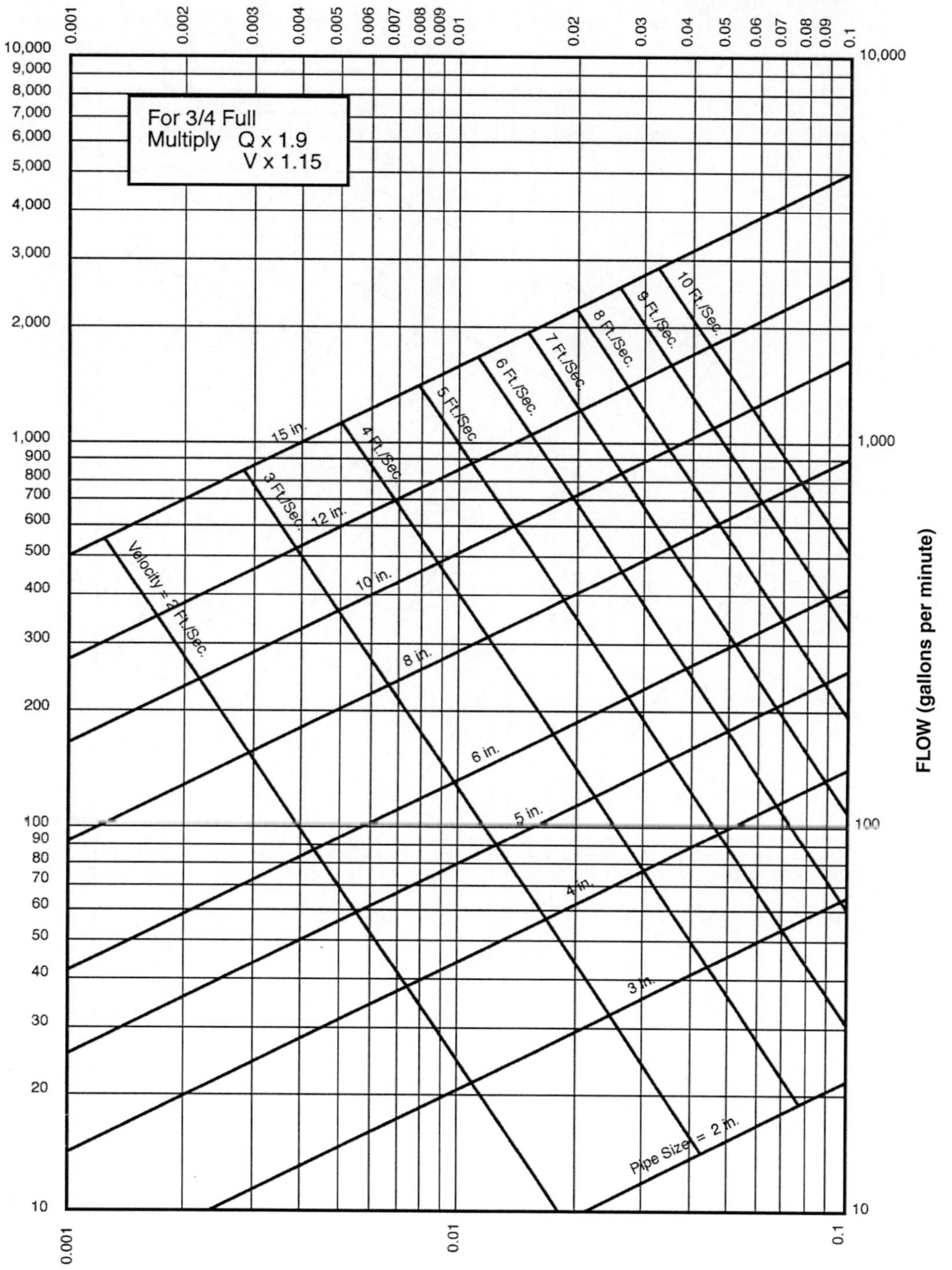

FLOW IN PARTLY FILLED (FULL) PIPES
(BASED ON MANNING'S FORMULA WITH n = .012)

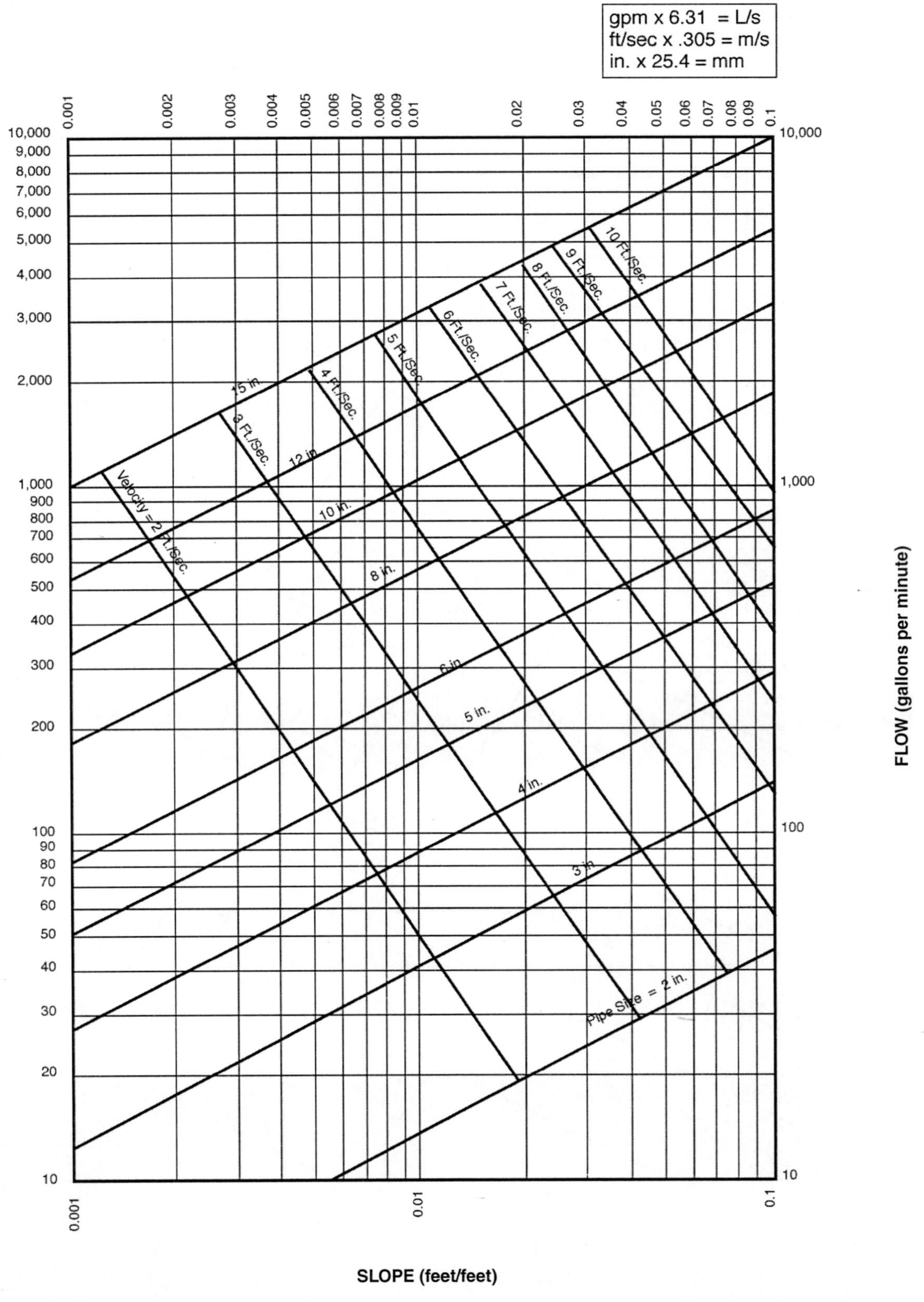

METRIC SYSTEM

(INTERNATIONAL SYSTEM OF UNITS – SI)

For the users of this code, we are including a short explanation and some conversion tables to aid in the conversion of our familiar English units to the forthcoming SI units.

This is written with the code users in mind, and will detail only those measurements used in everyday work and calculations.

GENERAL COMMENTS

Our present system of measuring involves the three dimensions of force, length. and time. The SI units involve mass, length, and time. The change of force to mass has meaning in scientific and engineering work, but for practical use in ordinary construction, we will show kilogram to pounds conversion values, although an exact conversion would be pounds force divided by the acceleration due to gravity to mass units.

In the same manner, the SI units for temperature expressed in Kelvins and based on absolute zero will be given as degrees Celsius, which is the more familiar and practical Centigrade degrees.

The SI system measures angles in radians where there are 2 pi radians in a circle, but using a 1.5708 bend to change from a vertical stack to a horizontal house drain is not as easy as calling out a ¼ bend or an ell for water piping.

The foregoing notes are intended to show that in making conversions from one unit system to another, a little common sense must be used and the degree of accuracy needed to do the job at hand.

The following tables are set up using this approach and using the preferred SI units.

UNIT CONVERSIONS

TO CONVERT	INTO	MULTIPLY BY
Atmosphere	Centimeters of mercury	76.0
British thermal units (Btu)	Joules	1055.056
Btus/hour	Kilowatts	0.000293
Btus/hour	Watts	0.293
Circumference	Radians	6.283
Cubic feet	Cubic meters	0.0283
Cubic feet	Liters	28.32
Cubic feet/hour	Cubic meters/hour	0.0283
Cubic feet/minute	Liters/second	0.4719
Cubic inches	Cubic meters	1.64×10^{-5}
Cubic inches	Liters	0.01639
Cubic meters	Gallons (U.S. liquid)	264.17
Cubic yards	Cubic meters	0.76455
Degrees	Radians	0.0175
Fahrenheit	Celsius	(ºF-32)/1.8
Feet	Meters	0.3048
Feet	Millimeters	304.8
Feet/second	Meters/second	0.3048
Foot-pounds	Joules	1.356
Foot pounds-force/minute	Kilowatts	2.260×10^{-5}
Foot pounds-force/second	Kilowatts	1.356×10^{-3}
Gallons	Liters	3.785
Gallons/day	Liters/second	4.3×10^{-5}
Gallons/minute	Liters/second	0.063
Grains	Kilograms	6.479×10^{-5}
Horsepower	Kilowatts	0.7457
Horsepower-hours	Joules	$2.684 \times 10^{+6}$
Horsepower-hours	Kilowatt-hours	0.7457
Inches	Millimeters	25.4
Inches/hour	Millimeters/hour	25.4
Inches of mercury (0ºC)	KiloPascals	3.3863
Joules	Btus	9.480×10^{-4}
Joules	Foot-pounds	0.7376
Joules	Watt-hours	2.778×10^{-4}
Kilograms	Pounds	2.2046
Kilograms	Tons (short)	1.102×10^{-3}
Kilometers	Miles (statute)	0.6214
Kilometers/hour	Miles/hour	0.6214
Kilowatts	Btus/hour	3412.14
Kilowatts	Horsepower	1.341
Kilowatt-hours	Btus	3413
Kilowatt-hours	Foot-pounds	$2.655 \times 10^{+6}$
Kilowatt-hours	Joules	$3.6 \times 10^{+6}$
Liters	Cubic feet	0.03531
Liters	Gallons (U.S. liquid)	0.2642

UNIT CONVERSIONS (continued)

TO CONVERT	INTO	MULTIPLY BY
Meters	Feet	3.281
Meters	Inches	39.37
Meters	Yards	1.094
Meters/second	Feet/second	3.281
Meters/second	Miles/hr	2.237
Miles (statute)	Kilometers	1.609
Miles/hour	Meters/minute	26.82
Millimeters	Inches	0.03937
Ounces (fluid)	Kilograms	0.02834
Pounds	Kilograms	0.4536
Pounds/foot	Kilograms/meters	1.4881
Pounds-force/square inch	KiloPascals	6.8947
Quarts (liquid)	Liters	0.9463
Radians	Degrees	57.30
Square feet	Square meters	0.0929
Square inches	Square meters	6.45×10^{-4}
Square inches	Square millimeters	645.16
Square meters	Square inches	1550
Square millimeters	Square inches	1.550×10^{-3}
Water column (1 inch)	KiloPascals	0.24908
Watts	Btus/hour	3.4121
Watts	Horsepower	1.341×10^{-3}

When the plumbing industry, including plumbers, suppliers, and manufacturers, actually begins the metric conversion program, it will undoubtedly follow the guidelines of committees selected from all phases of the construction industry as set up under the American National Metric Council.

The final preferred units used will be those that apply to our industry and will be of the magnitude to simplify and ease job calculations and avoid confusion and ambiguity.

The conversion looks complex and confusing, but when the metric system was first proposed in France, an attempt was made to include a ten-hour day, a ten-day week, and ten months to the year, but cooler heads prevailed and our time still follows the sun and seasons. Likewise, assigning new units or numbers to the quantities we must work with cannot change the basic hydraulic principles that plumbers have worked with throughout history.

Information on conversion factors is provided by ANSI, the American National Metric Council, and the Division of Designatronics, Inc.

TYPICAL PLUMBING FIXTURE SYMBOLS

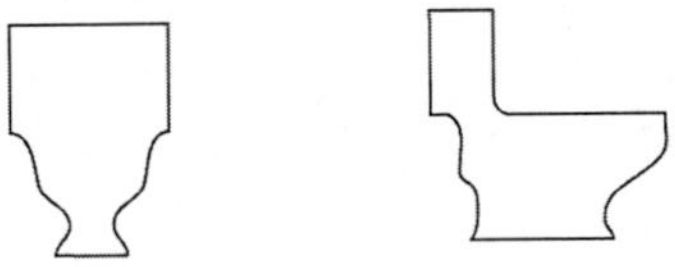

WATER CLOSETS
TANK TYPE
FLOOR MOUNTED

WATER CLOSETS
FLUSHOMETER TYPE
WALL MOUNTED

WATER CLOSETS
FLUSHOMETER TYPE
FLOOR MOUNTED

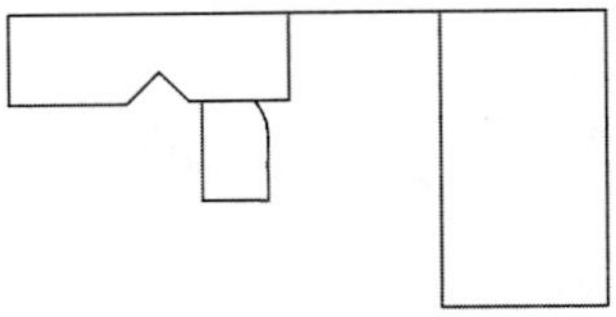

TWO PART SINK
GARBAGE DISPOSER
DISHWASHER

LAUNDRY TRAY

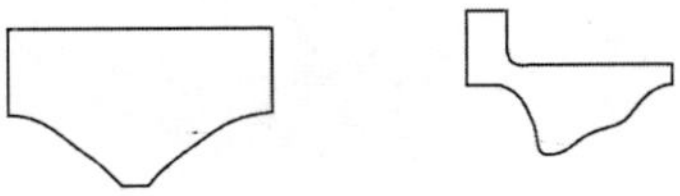

LAVATORY
(WASH BASIN)

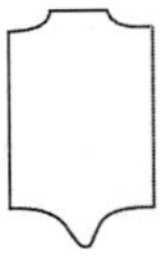

URINAL
WALL HUNG

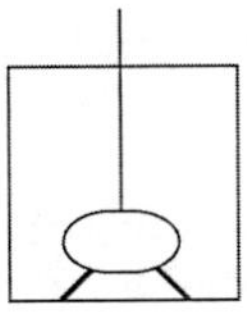

SUMP PIT
SEWAGE EJECTOR

STALL SHOWER

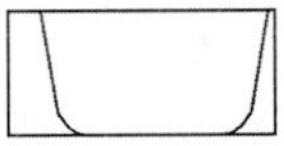

BATHTUB

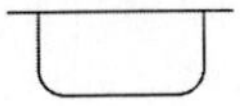

FLOOR SINK

FLOOR DRAIN

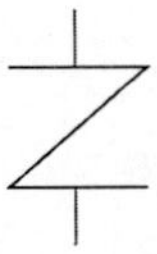

CHECK VALVE

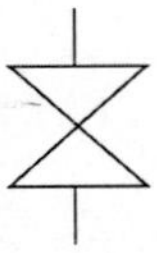

GATEVALVE

TRADE EXAMINATIONS
DRAINAGE AND VENTING SYSTEMS

Drawings #1 through #5 are elevation views of buildings showing locations of plumbing fixtures. Complete the drawings according to the following instructions:

1. Show sizes for all sections of drain piping, vent piping, traps, and cleanouts.

2. Do not oversize, use minimum sizes allowed by the Uniform Plumbing Code for unit loading on each drain pipe, vent pipe, and trap.

3. Check each building for the necessity of backwater valves, sewer ejectors, and sufficient vent area.

4. The example drawing is completed to show you what is required.

5. Use the tables in the code for sizing of drainage piping, vent piping, traps, and cleanouts.

6. Check your answers with the completed drawings in the answer section.

EXAMPLE DRAWING – DWV
Private Use

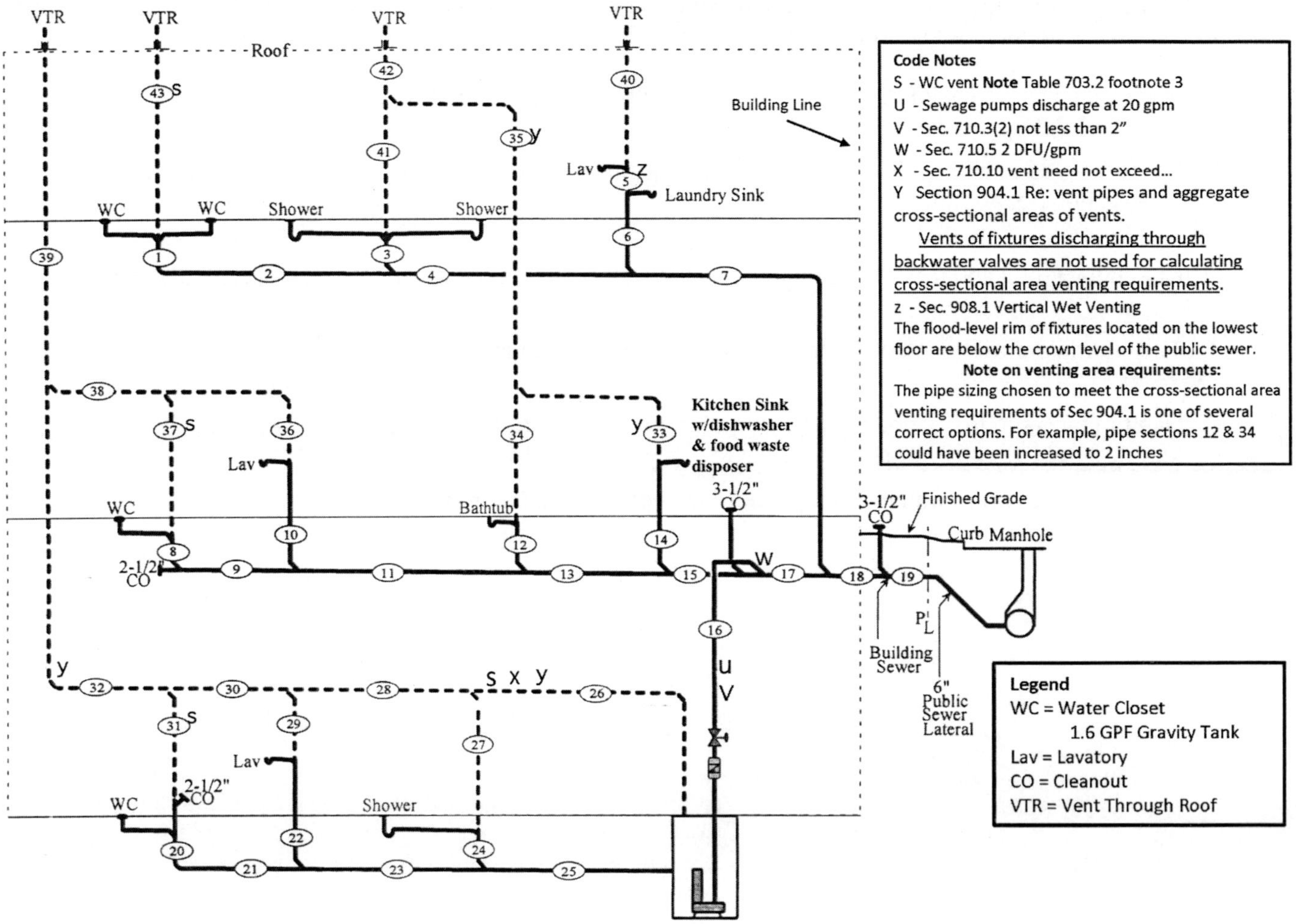

Pipe Section	Fixture Units	Pipe Size	Pipe Section	Fixture Units	Pipe Size	Pipe Section	Fixture Units	Pipe Size	Pipe Section	Fixture Units	Pipe Size	Pipe Section	Fixture Units	Pipe Size
1	6	3"	9	3	3"	17	48	4"	26	6	2"	35	4	2"
2	6	3"	10	1	1-1/4"	18	61	4"	27	2	1-1/2"	36	1	1-1/4"
3	4	2"	11	4	3"	19	61	4"	28	6	2"	37	3	2"
4	10	3"	12	2	1-1/2"	20	3	3"	29	1	1-1/4"	38	4	2"
5	1	2"	13	6	3"	21	3	3"	30	6	2"	39	10	2"
6	3	2"	14	2	2"	22	1	1-1/4"	31	3	2"	40	3	1-1/2"
7	13	3"	15	8	3"	23	4	3"	32	6	2"	41	4	1-1/2"
8	3	3"	16	20 gpm	2"	24	2	2"	33	2	2"	42	8	2"
						25	6	3"	34	2	1-1/2"	43	6	3"

DRAWING #1 –
Private Use

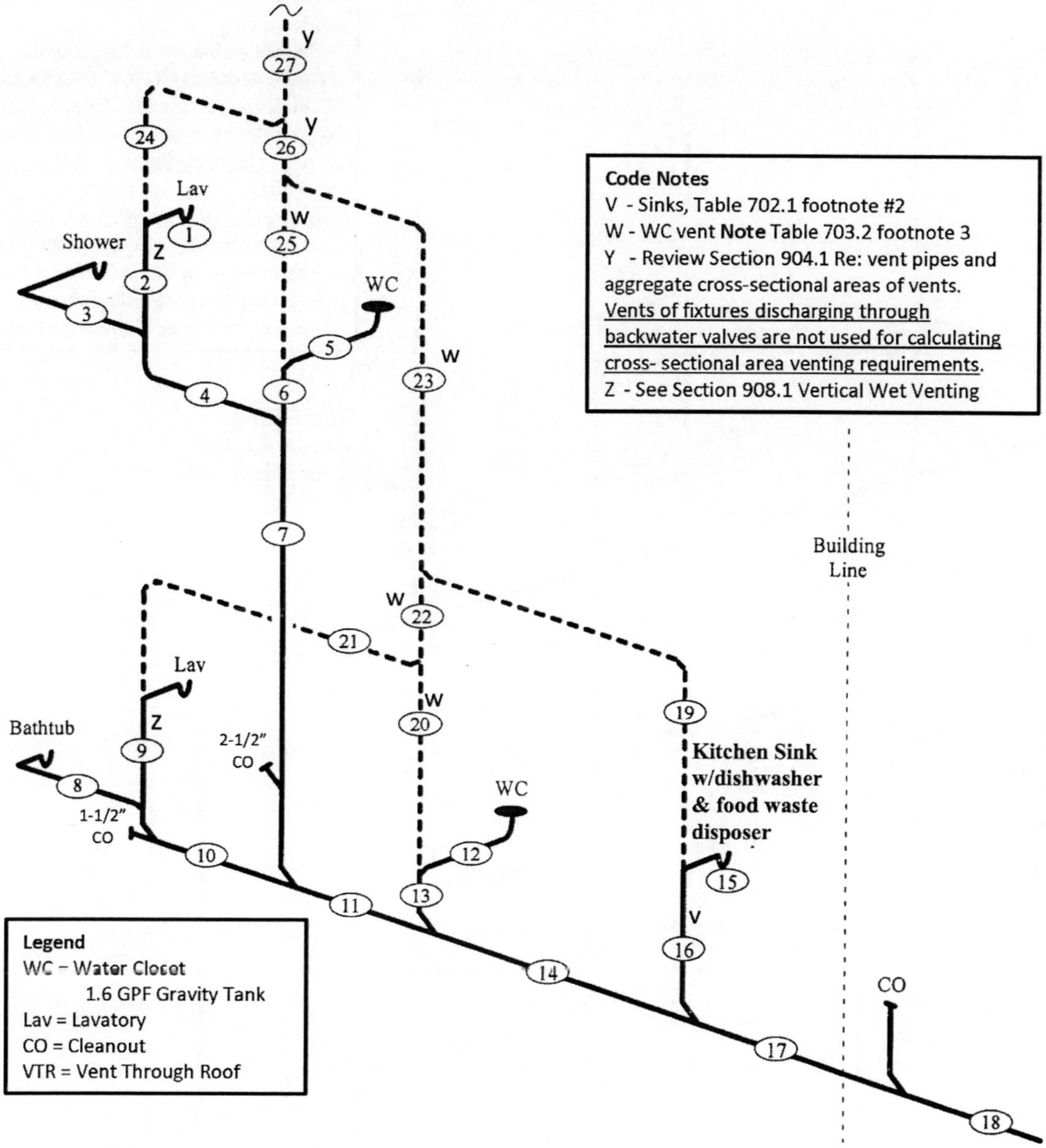

Code Notes
V - Sinks, Table 702.1 footnote #2
W - WC vent **Note** Table 703.2 footnote 3
Y - Review Section 904.1 Re: vent pipes and aggregate cross-sectional areas of vents. Vents of fixtures discharging through backwater valves are not used for calculating cross- sectional area venting requirements.
Z - See Section 908.1 Vertical Wet Venting

Legend
WC – Water Closet
1.6 GPF Gravity Tank
Lav = Lavatory
CO = Cleanout
VTR = Vent Through Roof

Pipe Section	Fixture Units	Pipe Size	Pipe Section	Fixture Units	Pipe Size	Pipe Section	Fixture Units	Pipe Size
1			10			19		
2			11			20		
3			12			21		
4			13			22		
5			14			23		
6			15			24		
7			16			25		
8			17			26		
9			18			27		

DRAWING #2
Private Use

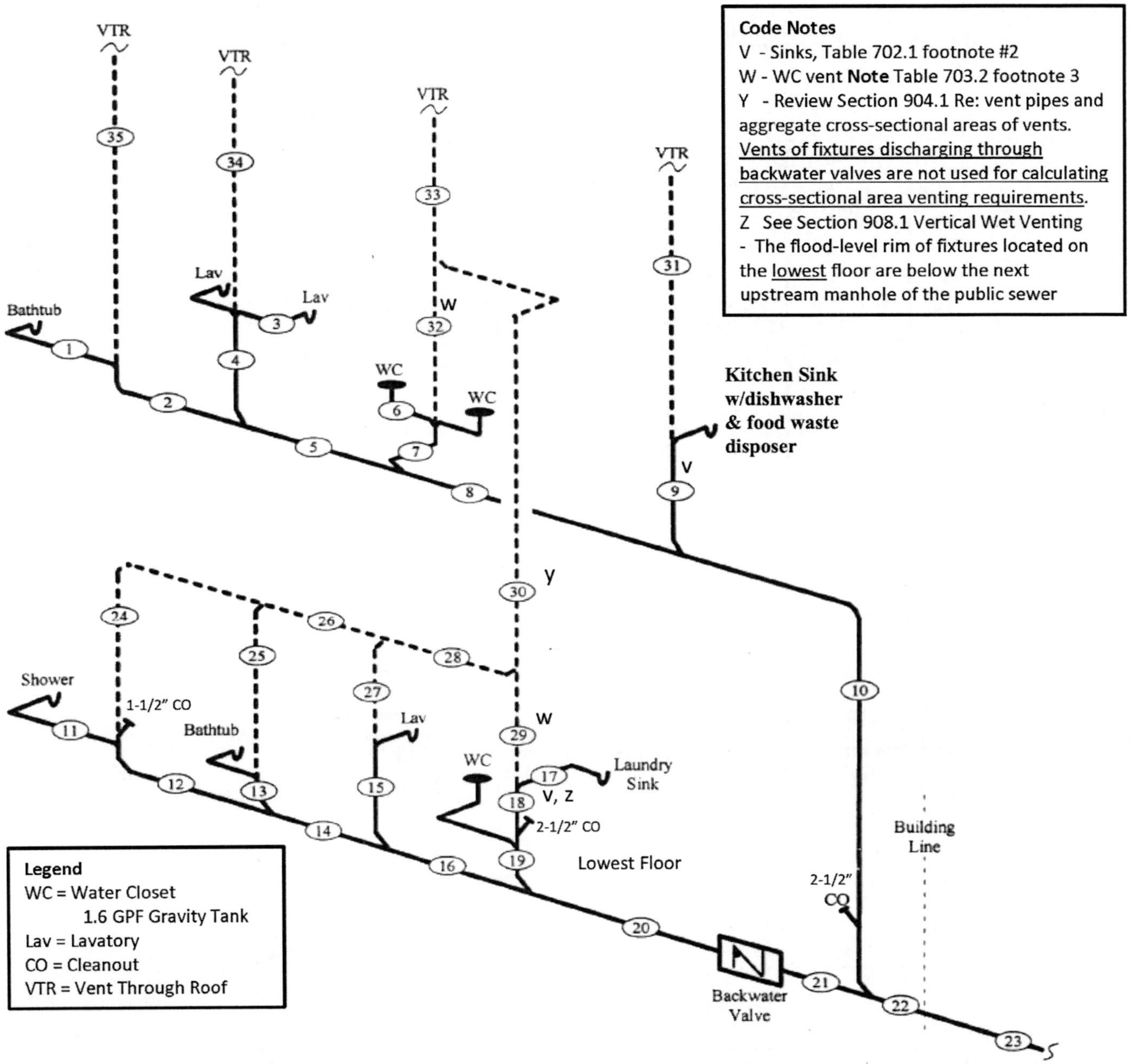

Pipe Section	Fixture Units	Pipe Size	Pipe Section	Fixture Units	Pipe Size	Pipe Section	Fixture Units	Pipe Size	Pipe Section	Fixture Units	Pipe Size	Pipe Section	Fixture Units	Pipe Size
1			8			15			22			29		
2			9			16			23			30		
3			10			17			24			31		
4			11			18			25			32		
5			12			19			26			33		
6			13			20			27			34		
7			14			21			28			35		

DRAWING #3
Public Use

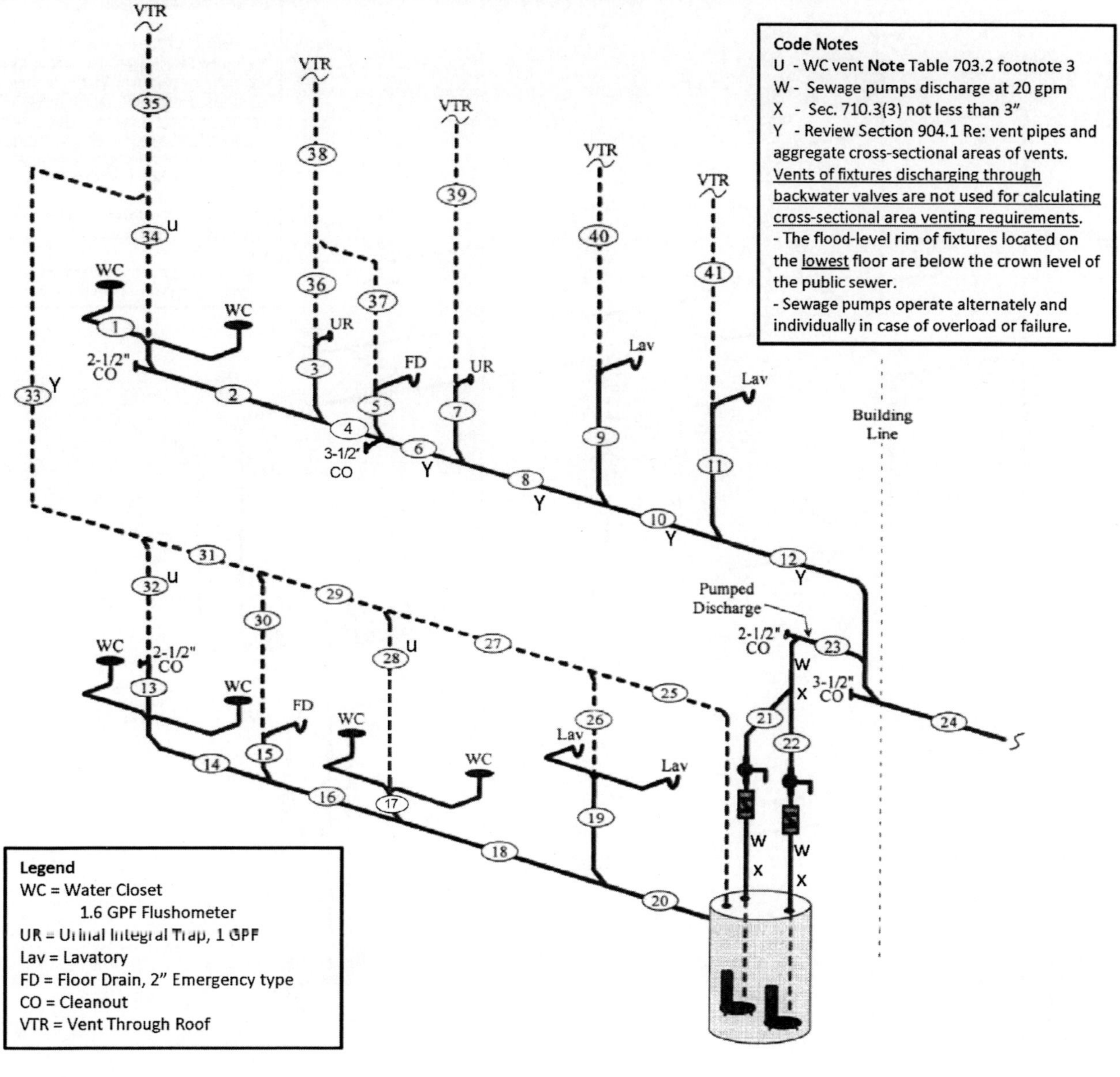

Pipe Section	Fixture Units	Pipe Size	Pipe Section	Fixture Units	Pipe Size	Pipe Section	Fixture Units	Pipe Size	Pipe Section	Fixture Units	Pipe Size	Pipe Section	Fixture Units	Pipe Size
1			9			17			25			33		
2			10			18			26			34		
3			11			19			27			35		
4			12			20			28			36		
5			13			21			29			37		
6			14			22			30			38		
7			15			23			31			39		
8			16			24			32			40		
												41		

DRAWING #4
Public Use

Code Notes
U - Note Table 703.2 footnote 3
W - Sewage pumps discharge at 20 gpm
X - Section 710.3(3) not less than 3"
Y - Review Section 904.1 Re: vent pipes and aggregate cross-sectional areas of vents. Vents from fixtures discharging through backwater valves are not used for calculating cross-sectional area venting requirements.
- The flood-level of fixtures on the middle floor are below next upstream manhole of the public sewer.
- The flood-level of fixtures located on the lowest floor are below crown level of the public sewer.
- Sewage pumps operate alternately and individually in case of overload or failure.

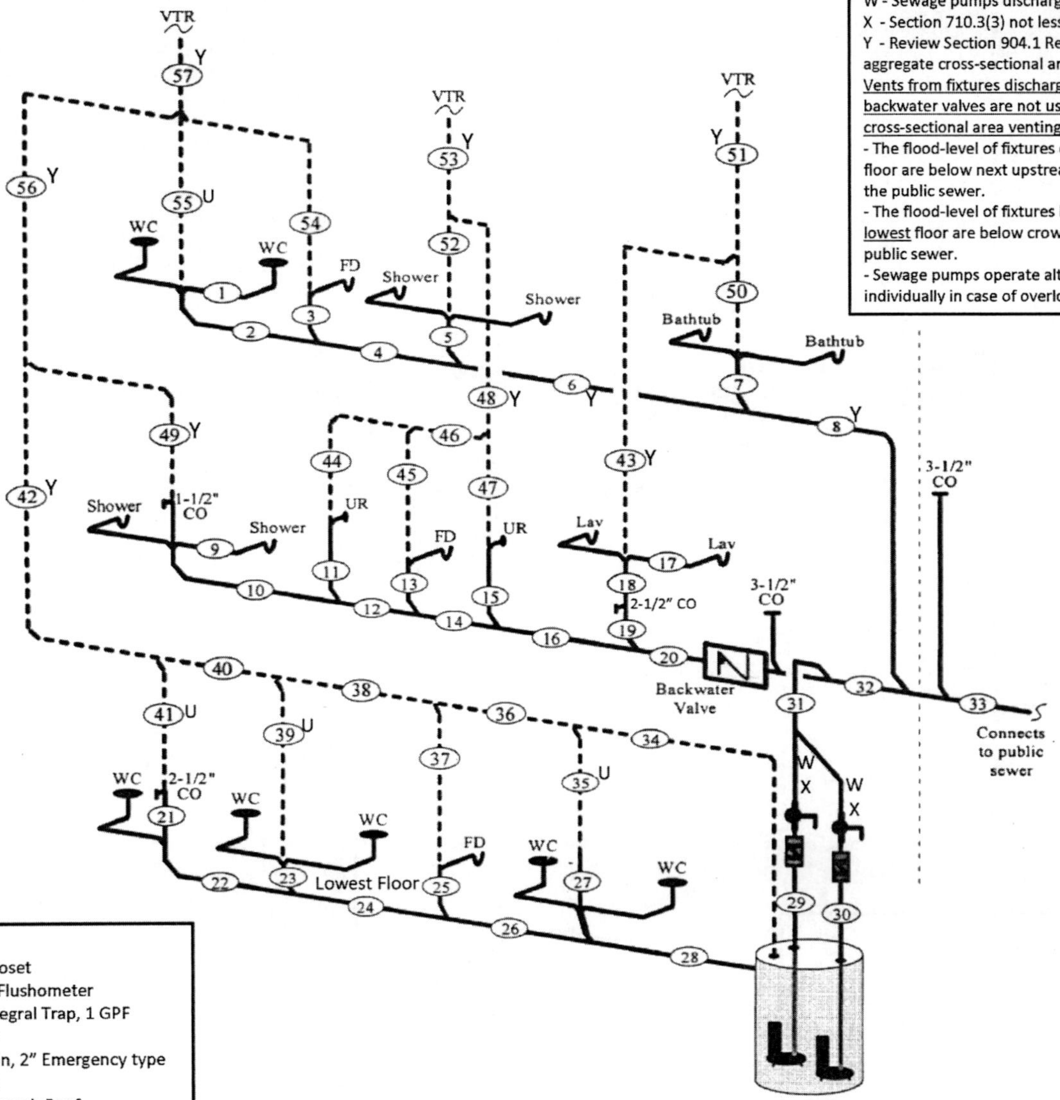

Legend
WC = Water Closet
1.6 GPF Flushometer
UR = Urinal Integral Trap, 1 GPF
Lav = Lavatory
FD = Floor Drain, 2" Emergency type
CO = Cleanout
VTR = Vent Through Roof

Pipe Section	Fixture Units	Pipe Size	Pipe Section	Fixture Units	Pipe Size	Pipe Section	Fixture Units	Pipe Size	Pipe Section	Fixture Units	Pipe Size	Pipe Section	Fixture Units	Pipe Size
1			13			25			36			47		
2			14			26			37			48		
3			15			27			38			49		
4			16			28			39			50		
5			17			29			40			51		
6			18			30			41			52		
7			19			31			42			53		
8			20			32			43			54		
9			21			33			44			55		
10			22			34			45			56		
11			23			35			46			57		
12			24											

DRAWING #5
Private Use

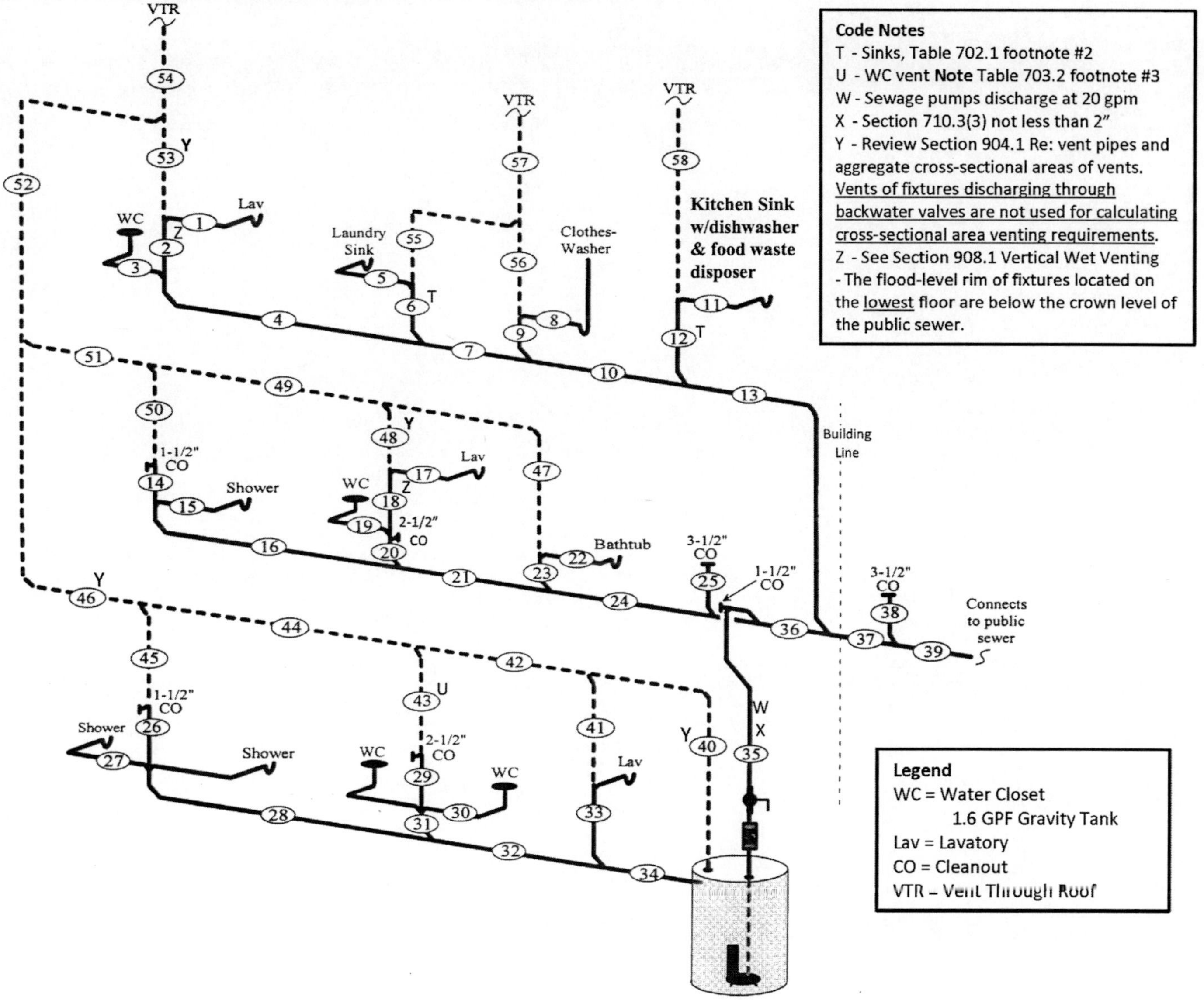

Pipe Section	Fixture Units	Pipe Size	Pipe Section	Fixture Units	Pipe Size	Pipe Section	Fixture Units	Pipe Size	Pipe Section	Fixture Units	Pipe Size	Pipe Section	Fixture Units	Pipe Size
1			13			25			37			48		
2			14			26			38			49		
3			15			27			39			50		
4			16			28			40			51		
5			17			29			41			52		
6			18			30			42			53		
7			19			31			43			54		
8			20			32			44			55		
9			21			33			45			56		
10			22			34			46			57		
11			23			35			47			58		
12			24			36								

TRADE EXAMINATIONS
WATER PIPING SYSTEMS

Drawings #6 through #11 are of water piping systems. Complete the drawing by sizing all sections of piping according to the following instructions:

1. The example drawing is completed to show what is required.

2. The information regarding developed length, low and high water pressure, and elevation is given for each drawing.

3. Construct a table as shown in the example for each drawing and enter the necessary information.

4. Cold water piping is shown in solid lines.

5. Hot water piping is shown in dashed lines.

6. Drawings #6 to #10 are water piping systems for the building in drawings #1 to #5, respectively.

7. Use the tables and procedures described in Chapter 6 of the Uniform Plumbing Code for sizing of cold and hot water piping.

8. Show sizes for all sections of piping.

9. Check your answers with the completed drawings in the answer section.

EXAMPLE DRAWING – Water
Private Use

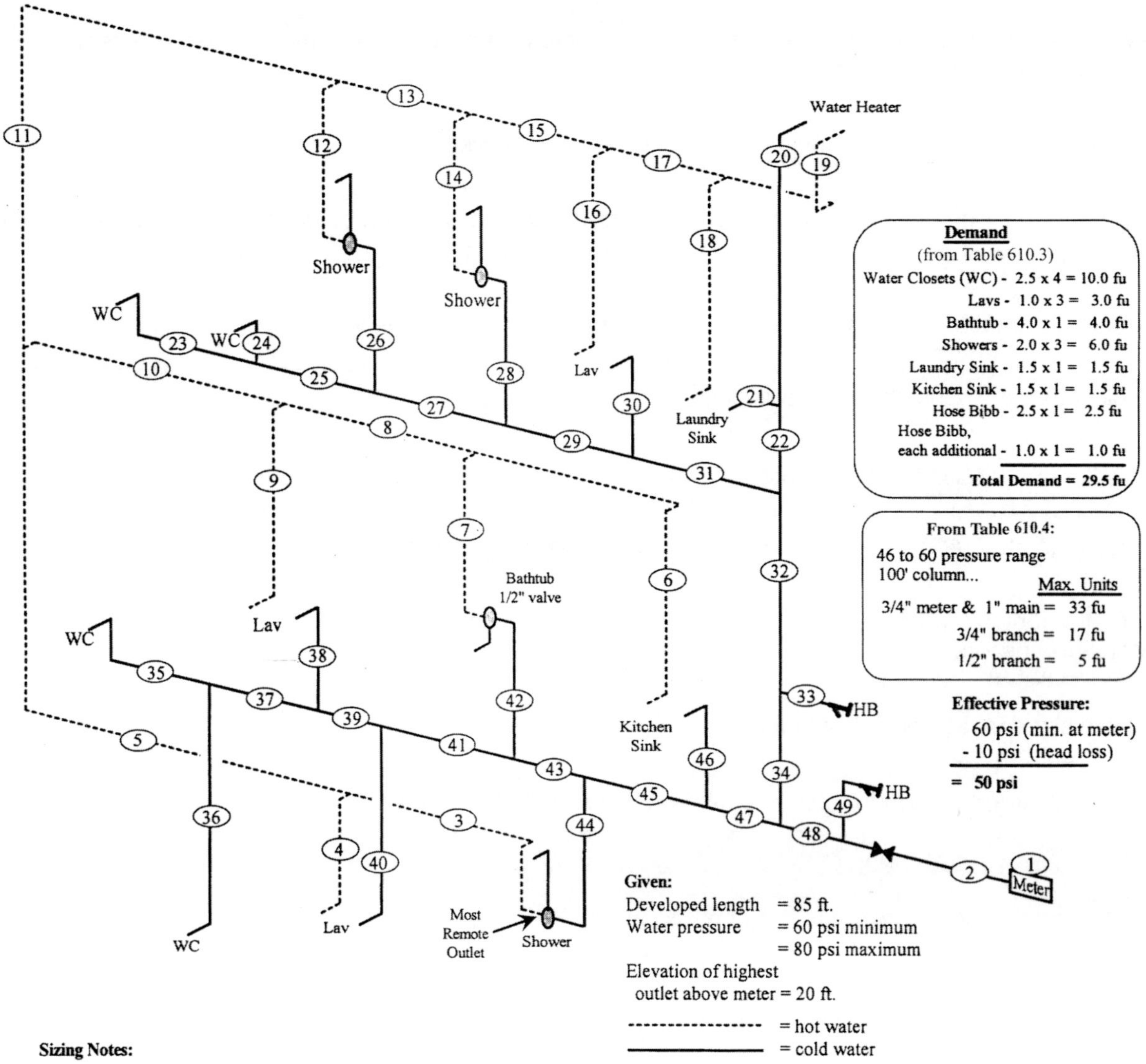

Sizing Notes:

All Water Closets (WC) are 1.6 GPF gravity tank

The developed length is determined by measuring the distance from the most remote water outlet back to the meter or other source of supply. In this drawing, the most remote outlet is the hot water branch serving the shower in the lower left hand corner. The most remote outlet may be either a hot or cold water outlet.

Use caution when sizing pipe sections serving "hose bibbs" and additional "hose bibbs." Only one hose bibb served by any given pipe section is assigned a fixture unit value of 2.5. All others served by the same pipe section are assigned a fixture unit value of 1.0, hence the term "additional."

When sizing water distribution systems containing hot water, the total fixture unit value for the hot water system is added to the cold water demand as you work your way back toward the water meter or other source of supply. You may eventually reach a point where the combined hot and cold demand could exceed the originally established demand for the building. Section 610.9 states: Where Table 610.4 is used, the minimum size of each branch shall be determined by the total fixture units served by that branch and then following the steps in section 610.8. No branch piping shall exceed the total demand in fixture units for the system computed from table 610.3 (see the asterisk in the table).

Pipe Section	Fixture Units	Pipe Size	Pipe Section	Fixture Units	Pipe Size	Pipe Section	Fixture Units	Pipe Size
1-Meter	29.5	3/4"	17	14.5	3/4"	33	2.5	1/2"
2	29.5	1"	18	1.5	1/2"	34	29.5*	1"
3	2.0	1/2"	19	16.0	3/4"	35	2.5	1/2"
4	1.0	1/2"	20	16.0	3/4"	36	2.5	1/2"
5	3.0	1/2"	21	1.5	1/2"	37	5.0	1/2"
6	1.5	1/2"	22	17.5	1"	38	1.0	1/2"
7	4.0	1/2"	23	2.5	1/2"	39	6.0	3/4"
8	5.5	3/4"	24	2.5	1/2"	40	1.0	1/2"
9	1.0	1/2"	25	5.0	1/2"	41	7.0	3/4"
10	6.5	3/4"	26	2.0	1/2"	42	4.0	1/2"
11	9.5	3/4"	27	7.0	3/4"	43	11.0	3/4"
12	2.0	1/2"	28	2.0	1/2"	44	2.0	1/2"
13	11.5	3/4"	29	9.0	3/4"	45	13.0	3/4"
14	2.0	1/2"	30	1.0	1/2"	46	1.5	1/2"
15	13.5	3/4"	31	10.0	3/4"	47	14.5	3/4"
16	1	1/2"	32	27.5	1"	48	29.5*	1"
						49	2.5	1/2"

DRAWING #6
Private Use

Given:
Developed Length = 95 ft
PSI at the meter = 50 psi min.
= 78 psi max.
Elevation Change + 30 ft
Cold water = ————
Hot water = ··········

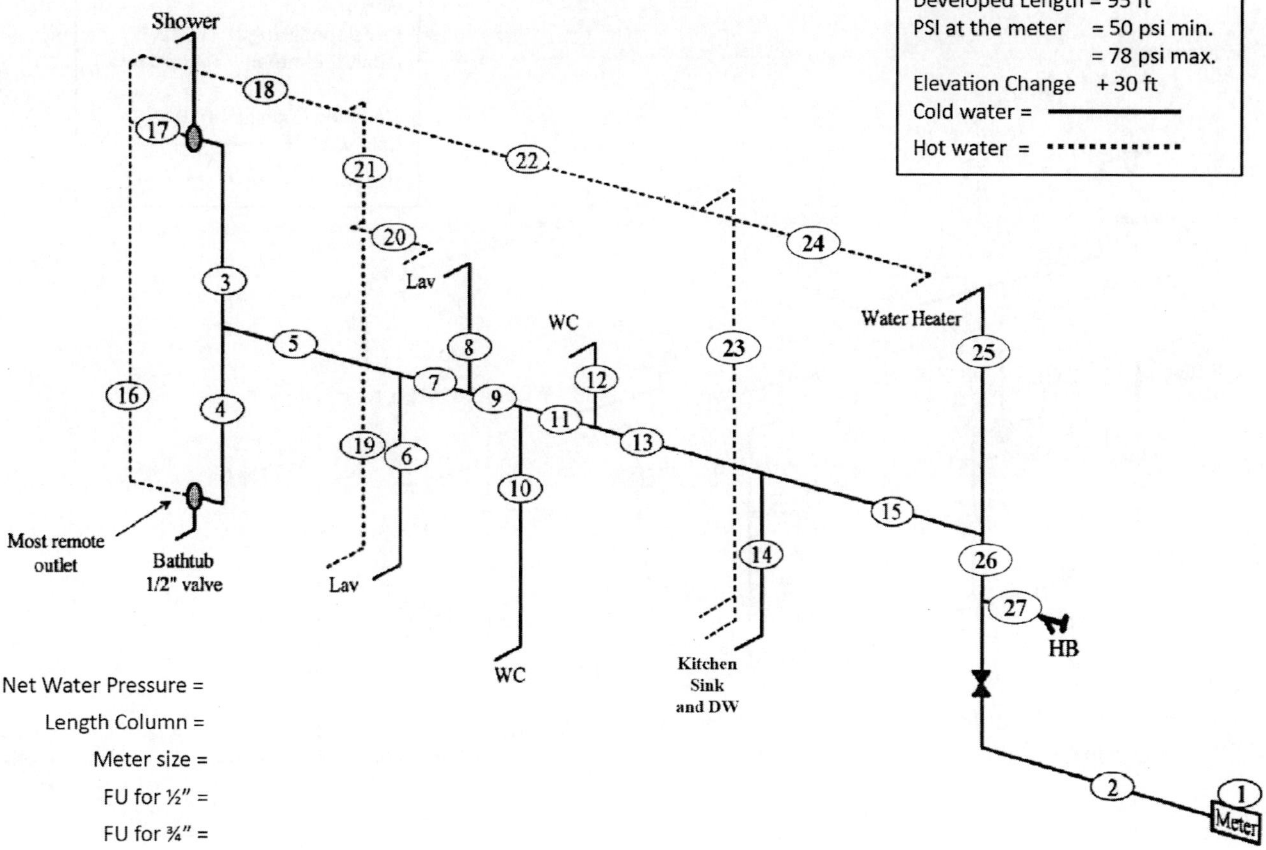

Net Water Pressure =

Length Column =

Meter size =

FU for ½" =

FU for ¾" =

FU for 1" =

Demand from Table 610.3 in WSFU

Fixture	Building Supply WSFU	QTY	Total WSFU	Branches Cold FU @100%	Hot FU @100%
WC, 1.6 GPF (Gravity Tank)					
Lavatory					
Bathtub					
Shower					
Kitchen Sink with DW					
Laundry Sink					
Clothes Washer					
Bar Sink					
Hose Bibb					
Additional Hose Bibbs					
Wash Fountain, Circular Sink					
Total Demand at Meter					
Total for Hot and Cold Demands					

	FU	SIZE		FU	SIZE
1			16		
2			17		
3			18		
4			19		
5			20		
6			21		
7			22		
8			23		
9			24		
10			25		
11			26		
12			27		
13					
14					
15					

DRAWING #7
Private Use

Given:
Developed Length = 81 ft
PSI at the meter = 48 psi min.
= 75 psi max.
Elevation Change + 10 ft
Cold water = ————
Hot water = ▪▪▪▪▪▪▪▪▪▪▪▪

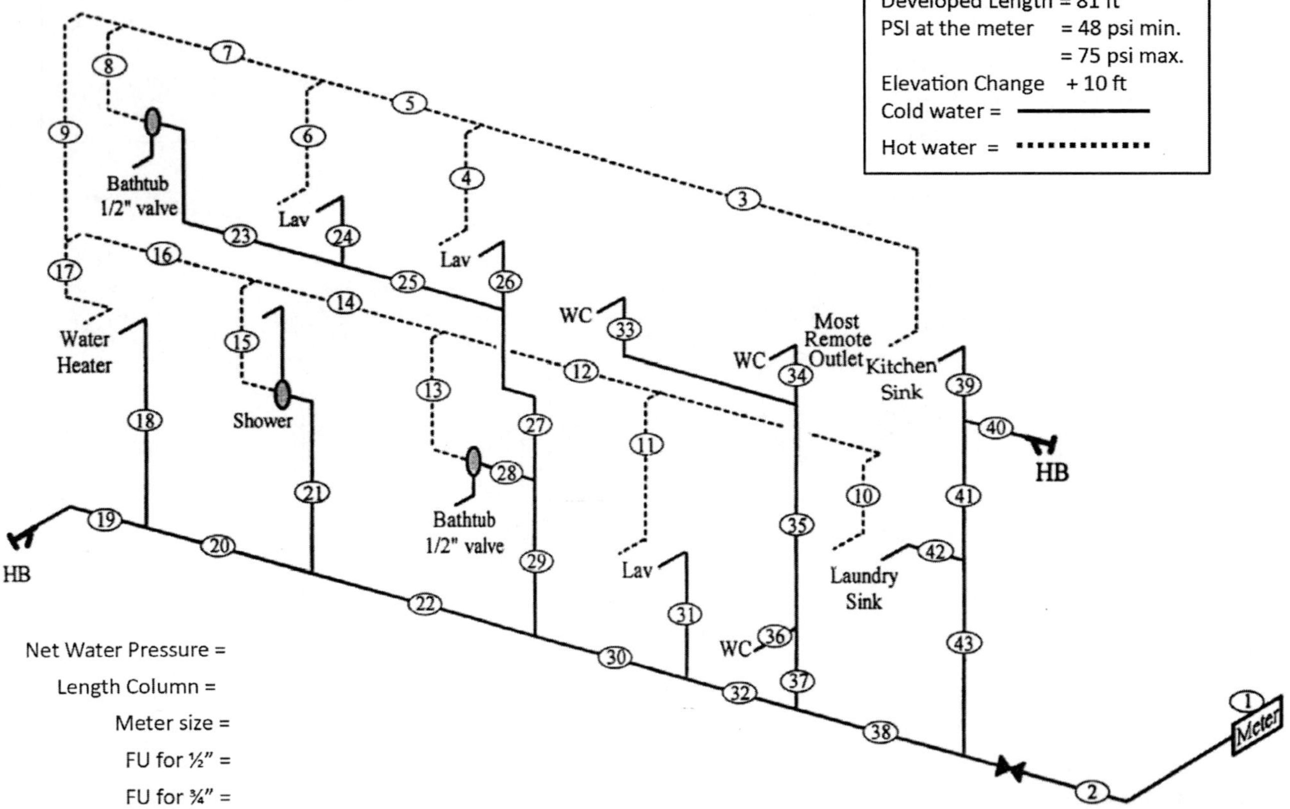

Net Water Pressure =

Length Column =

Meter size =

FU for ½" =

FU for ¾" =

FU for 1" =

FU for 1-1/4" =

Demand from Table 610.3 in WSFU

Fixture	Building Supply: WSFU	Building Supply: QTY	Building Supply: Total WSFU	Branches: Cold FU @100%	Branches: Hot FU @100%
WC, 1.6 GPF (Gravity Tank)					
Lavatory					
Bathtub					
Shower					
Kitchen Sink with DW					
Laundry Sink					
Clothes Washer					
Bar Sink					
Hose Bibb					
Additional Hose Bibbs					
Wash Fountain, Circular Sink					
Total Demand at Meter					
Total for Hot and Cold Demands					

	FU	SIZE		FU	SIZE		FU	SIZE
1			16			31		
2			17			32		
3			18			33		
4			19			34		
5			20			35		
6			21			36		
7			22			37		
8			23			38		
9			24			39		
10			25			40		
11			26			41		
12			27			42		
13			28			43		
14			29					
15			30					

DRAWING #8
Public Use

Net Water Pressure =
Length Column =
Meter size =
FU for ½" =
FU for ¾" =
FU for 1" =
FU for 1-1/4" =

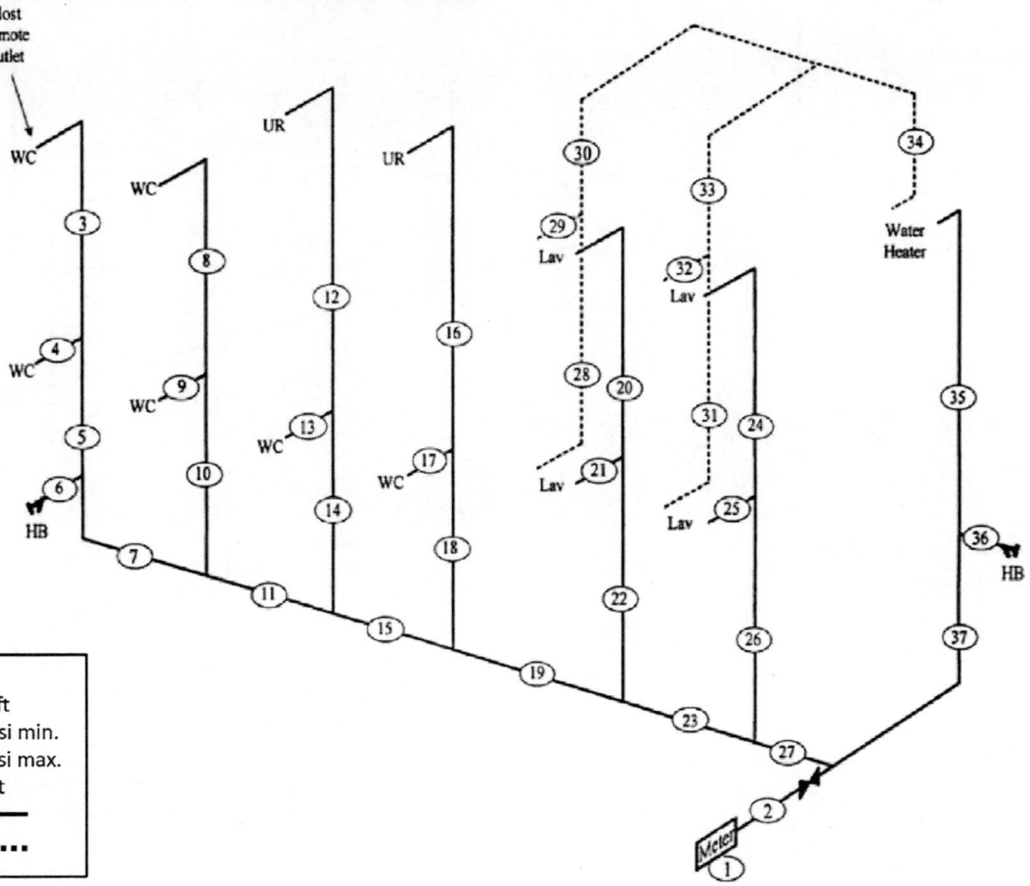

Given:
Developed Length = 100 ft
PSI at the meter = 62 psi min.
= 75 psi max.
Elevation Change + 24 ft
Cold water = ————
Hot water = ··············

Demand from Table 610.3 in WSFU

	Building Supply			Branches	
Fixture	**WSFU**	**QTY**	**Total WSFU**	**Cold FU @100%**	**Hot FU @100%**
WC, 1.6 GPF (Flushometer Tank)					
Urinal 1.0 GPF (Flushometer)*					
Urinal 1.0 GPF (Flushometer)*					
* See Footnote 7					
Lavatory					
Hose Bibb					
Additional Hose Bibbs					
Wash Fountain, Circular Sink					
Total Demand at Meter					
Total for Hot and Cold Demands					

	FU	SIZE		FU	SIZE
1			21		
2			22		
3			23		
4			24		
5			25		
6			26		
7			27		
8			28		
9			29		
10			30		
11			31		
12			32		
13			33		
14			34		
15			35		
16			36		
17			37		
18					
19					
20					

DRAWING #9
Public Use

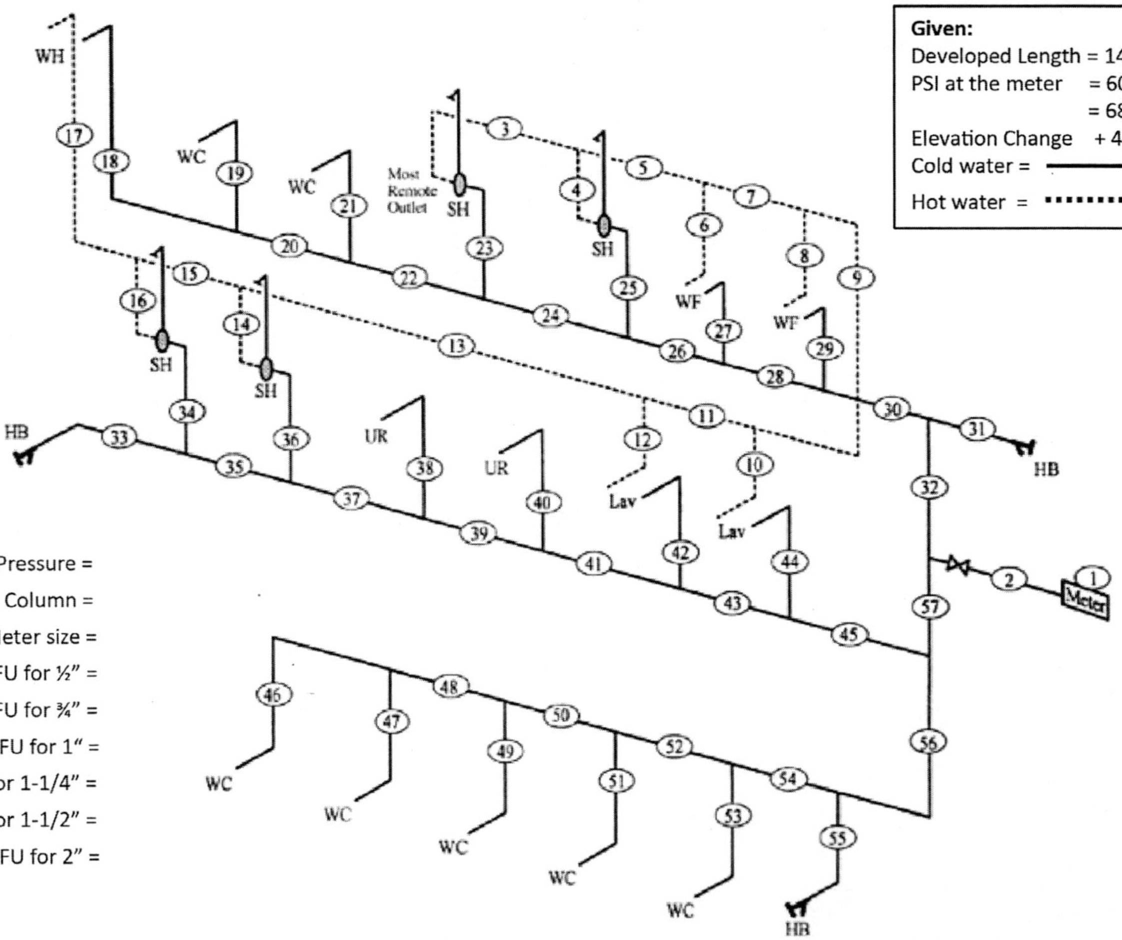

Given:
Developed Length = 145 ft
PSI at the meter = 60 psi min.
= 68 psi max.
Elevation Change + 40 ft
Cold water = ————
Hot water = ▪▪▪▪▪▪▪▪▪▪▪▪

Net Water Pressure =
Length Column =
Meter size =
FU for ½" =
FU for ¾" =
FU for 1" =
FU for 1-1/4" =
FU for 1-1/2" =
FU for 2" =

Demand from Table 610.3 in WSFU

Fixture	Building Supply			Branches	
	WSFU	QTY	Total WSFU	Cold FU @100%	Hot FU @100%
WC, 1.6 GPF (Flushometer)*					
WC, 1.6 GPF (Flushometer)*					
WC, 1.6 GPF (Flushometer)*					
WC, 1.6 GPF (Flushometer)*					
WC, 1.6 GPF (Flushometer)*					
Urinal 1.0 GPF (Flushometer)*					
Urinal 1.0 GPF (Flushometer)*					
* See Footnote 7					
Lavatory					
Shower					
Hose Bibb					
Additional Hose Bibbs					
Wash Fountain, Circular Spray					
Total Demand at Meter					
Total for Hot and Cold Demands					

	FU	SIZE		FU	SIZE		FU	SIZE
1			21			41		
2			22			42		
3			23			43		
4			24			44		
5			25			45		
6			26			46		
7			27			47		
8			28			48		
9			29			49		
10			30			50		
11			31			51		
12			32			52		
13			33			53		
14			34			54		
15			35			55		
16			36			56		
17			37			57		
18			38					
19			39					
20			40					

DRAWING #10
Private Use

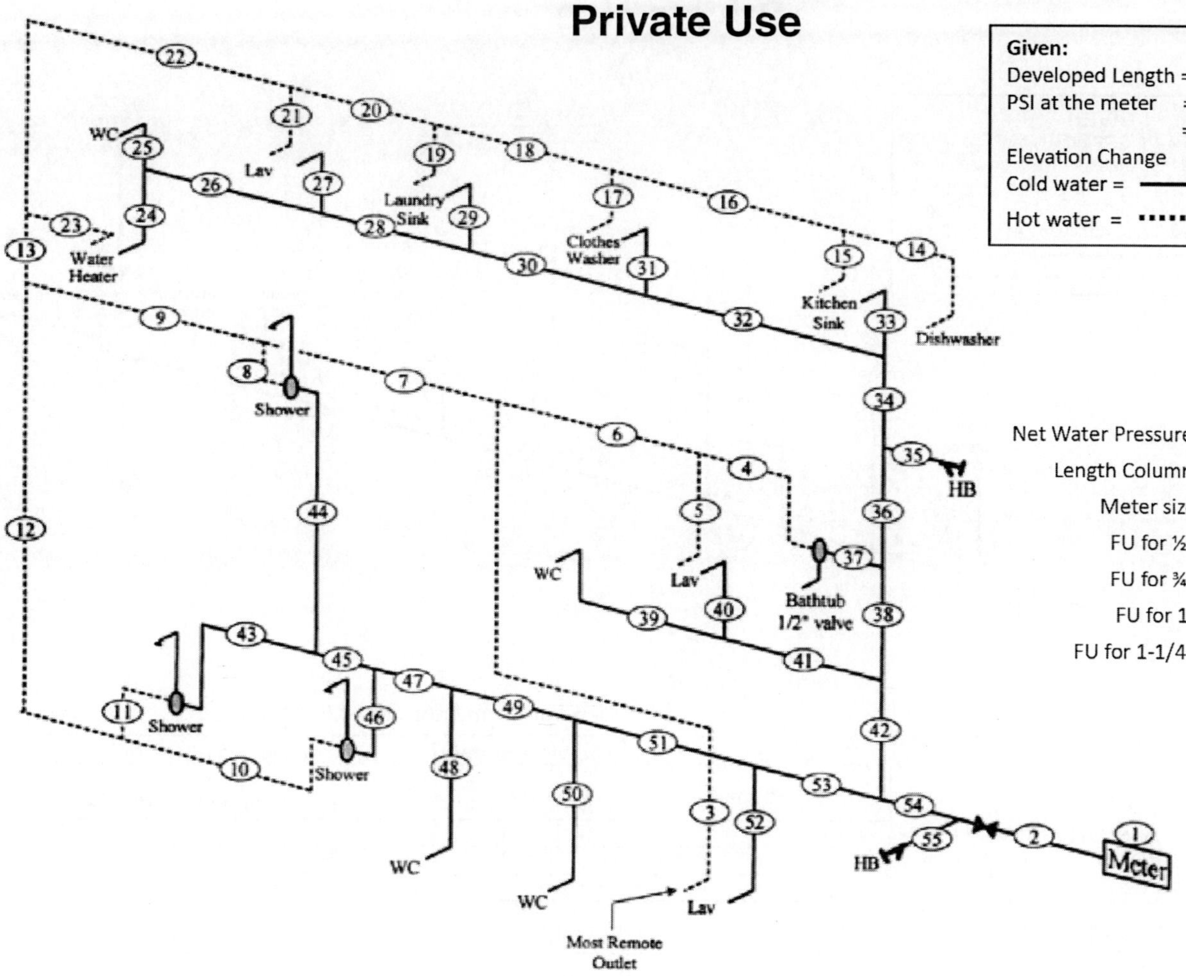

Given:
Developed Length = 90 ft
PSI at the meter = 75 psi min.
= 80 psi max.
Elevation Change + 30 ft
Cold water = ————
Hot water = ············

Net Water Pressure =
Length Column =
Meter size =
FU for ½" =
FU for ¾" =
FU for 1" =
FU for 1-1/4" =

Demand from Table 610.3 in WSFU

Fixture	Building Supply WSFU	QTY	Total WSFU	Branches Cold FU @100%	Hot FU @100%
WC, 1.6 GPF (Gravity Tank)					
Lavatory					
Bathtub					
Shower					
Kitchen Sink with DW					
Laundry Sink					
Clothes Washer					
Bar Sink					
Hose Bibb					
Additional Hose Bibbs					
Wash Fountain, Circular Sink					
Total Demand at Meter					
Total for Hot and Cold Demands					

	FU	SIZE		FU	SIZE		FU	SIZE
1			21			41		
2			22			42		
3			23			43		
4			24			44		
5			25			45		
6			26			46		
7			27			47		
8			28			48		
9			29			49		
10			30			50		
11			31			51		
12			32			52		
13			33			53		
14			34			54		
15			35			55		
16			36					
17			37					
18			38					
19			39					
20			40					

DRAWING #11
Private Use

GIVEN	
All water closets are 1.6 GPF gravity tank	
Developed Length from meter to most remote outlet.	
(Hot water at bathtub)	= 165 ft.
Minimum pressure at meter	= 62 psi
Elevation of highest fixture above meter	= 12 ft.
———————	= Cold
- - - - - - - - -	= Hot

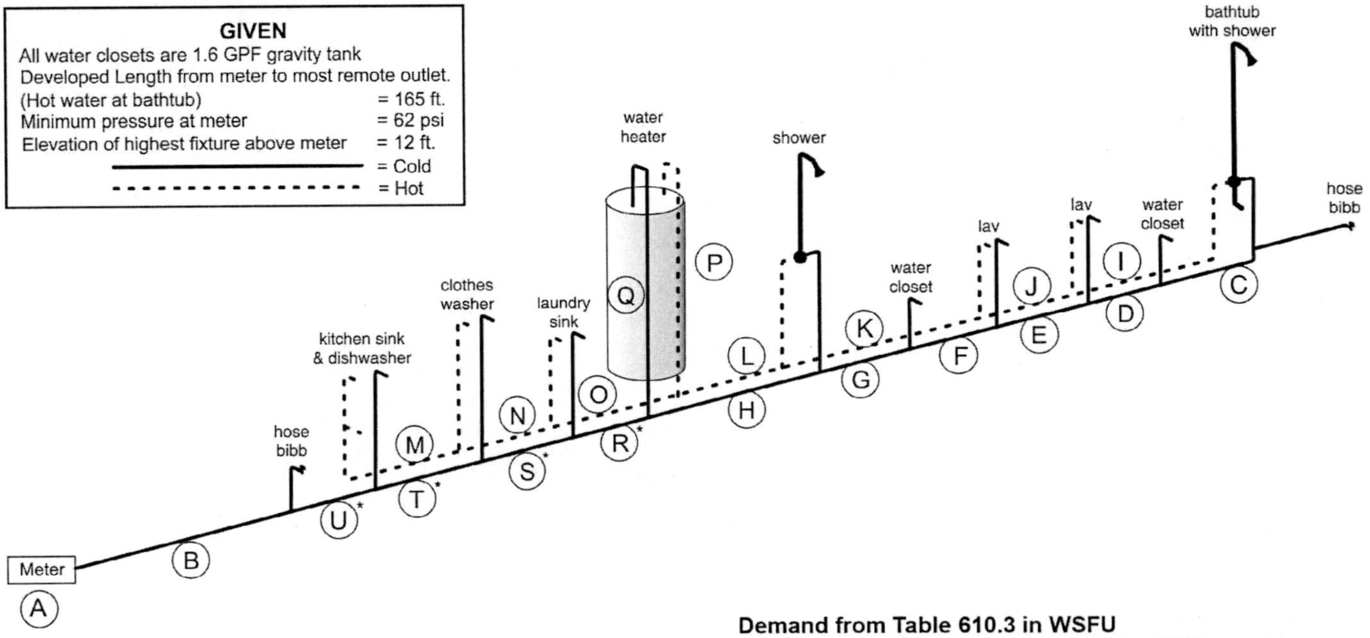

From Table 610.4
Pressure Range - 46 to 60 psi 200 foot column

Pipe Size	Max. Fixture Units
1"	
3/4"	
1/2"	

Demand from Table 610.3 in WSFU

	Building Supply			Branches - 75% Method		
Fixture	WSFU	QTY	Total WSFU	Cold WSFU (each outlet)	Hot WSFU (each outlet)	Total Hot WSFU
WC, 1.6 gpf						
Lavatory						
Shower						
Combination Bath/Shower						
Laundry Sink						
Clothes Washer						
Kitchen sink w/Dishwasher						
1st. Hose Bibb						
Additional Hose Bibbs						
Total Demand at water meter						
Total Demand at water heater						

Hose bibb sizing. Each pipe section serving only one hose bibb is assigned a fixture unit value of 2.5. Section B serves on hose bibb with a rating of 2.5 plus one "additional" hose bibb with a fixture unit value 1.0. The total hose bibb fixture unit value at section B is 3.5. See Note 8 at bottom of the Table 610.3.

Pipe Section	Fixture Units	Pipe Size
A meter		
B		
C		
D		
E		
F		
G		
H		
I		
J		
K		

Pipe Section	Fixture Units	Pipe Size
L		
M		
N		
O		
P		
Q		
R		
S		
T		
U		

*NOTE: When the demand at any point in the water distribution system would exceed the "total demand" you stop adding fixture units and maintain the total demand count to the meter. Section 610.9 states: "No branch piping shall exceed the total demand in fixture units for the system computed from Table 610.3."

TRADE EXAMINATION
GAS PIPING SYSTEMS

Drawings #12 through #17 are of natural fuel-gas systems. Drawing #17 is an undiluted propane system. Complete the drawings by sizing all sections of piping using the method specified in the drawings (See Section 1215.1).

1. The example drawing is given to illustrate what is required for a completed gas piping system. The answers are given for both the "Longest Length" and the "Branch Length" methods. Please note that the "Branch Length" method may result in pipe section sizes being smaller than those obtained using the "Longest Length" method.

2. Natural gas sizing tables are based on a specified piping or tubing material and the use of natural gas having a specific gravity of 0.60, supplied with an inlet pressure from less than 2.0 psi or less to 5.0 psi. Undiluted propage regulations and tables are based on the use of gas having a specific gravity of 1.50 and an inlet pressure varying from 0.4 psi to 10.0 psi.

3. The hourly volume of gas required at each piping outlet shall be taken as not less than the maximum hourly demand in Btu/h specified by the manufacturer of the appliance connected to each such outlet.

4. For Drawings #12 through #17, some appliances have been assigned specific Btu input. For appliances that have not been assigned specific Btu input, refer to the approximate values shown in Table 1208.3.1 and size the drawing accordingly.

5. To obtain the cubic feet per hour of natural gas required at each outlet, divide the Btu input of the appliance by the average Btu heating value per cubic foot of the gas. The Btu heating value per cubic foot of gas can be obtained from the serving gas supplier or by finding the 'therm multiplier' on your gas bill and multiplying by 1,000. In the following examinations, use the specific heating value provided in the Given box.

6. Figure 1215.1.1 "Example Illustrating Use of Table 1208.3.1 and Table 1215.2(1)" in the Uniform Plumbing Code should be studied and understood. Note: the heating value of 1,100 btu/cu ft is **only an example** used to demonstrate the conversion process.

7. Check your answers with the completed drawings in the answer section.

Examples of Gas Pipe Sizing

Using "Longest Length" & "Branch Length" methods

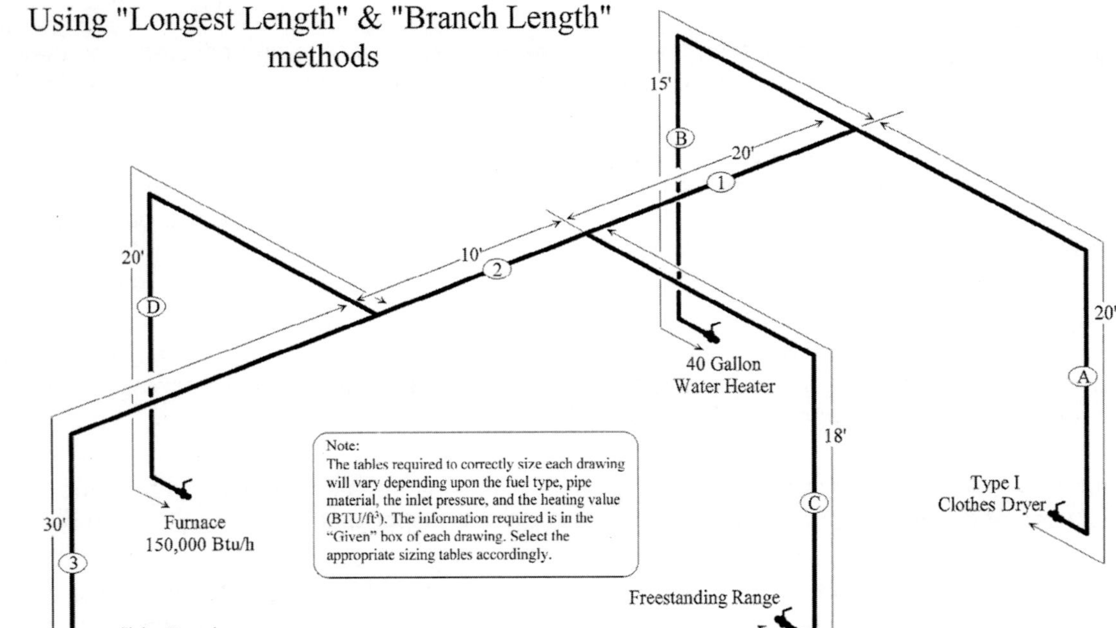

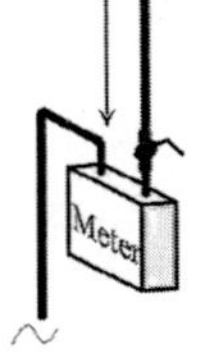

Sizing Procedure:

1) The Btu/h input for each gas appliance was obtained from Table 1208.3.1 or from the rating plate on the appliance. If the fuel is natural gas, then the Btu/h rating must be converted to cubic feet per hour (CFH). Divide the input rating of the gas appliance (Btu/h) by the heating value for each cubic foot of gas (Btu/ft³) supplied in the Given box for each drawing. For example, if the Btu/h rating of an appliance is 60,000 and the heating value per cubic foot of gas is 1,050, then divide 60,000 by 1,050 to obtain 57.14 CFH. Then round to the nearest whole number (round up when 0.5 or greater) to yield 57 CFH.
2) Determine the length from the meter to the most remote appliance in the system. In this example the clothes dryer is 80' from the meter.
3) When using the "longest length method" (Section 1215.1.1) all the pipe sizing must be taken from the 80' row of Table 1215.2(1).

3a) Size pipe section A using the information above (80' row of Table 1215.2(1).
Size pipe sections 1, 2 and 3 from the same 80' row of table 1215.2(1) based on the CFH demand of that section which will increase as you work back toward the meter or supply source.

3b) Size the remaining branches B, C and D based on their individual demand using the same row and table.

Definitions for Sizing Purposes:

Pipe section — A section of pipe which is located:

1. Between the meter and the first branch(s)
2. Between a branch and the next branch
3. Between a branch and the appliance outlet

Section - Two or more pieces joined by a coupling, 90° or 45° elbow are considered part of one section.

Outlet - The pipe section located between a branch and an appliance stub out.

Branch Length Method (Section 1215.1.2)
Every section supplying the most remote outlet is sized in the same column. This example is 80'.
The remaining branches are sized in the column (or next longer column) represented by the distance from the source to most remote outlet.
Section B 75' from meter, uses the 80' row
Section C 58' from meter, uses the 60' row
Section D 50' from meter, uses the 50' row

In this example each cubit foot of gas contains 1050 BTU's. (From gas supplier)
To calculate CFH demand divide BTUs/H by 1050 BTUs/Cu. Ft.

From Table 1215.2(1)

Pipe Size	1/2"	3/4"	1"	1-1/4"	
80' row	56	117	220	452	CFH

Pipe Section / Outlet	Appliance Served	BTU/H Table 1208.4.1	CFH Demand	Longest Length Method: Length Table 1215.2(1)	Longest Length Method: Pipe Size Table 1215.2(1)	Branch Length Method: Length Table 1215.2(1)	Branch Length Method: Pipe Size Table 1215.2(1)	
A	Type I Clothes Dryer	35,000	33	80'	1/2"	80'	1/2"	
B	40 Gallon Water Heater	35,000	33	80'	1/2"	80'	1/2"	
C	Freestanding Range	65,000	62	**80'**	**3/4"**	**60'**	**1/2"**	←Using 60' row
D	Furnace	150,000	143	**80'**	**1"**	**50'**	**3/4"**	←Using 50' row
1	Pipe Sections A & B	70,000	67	80'	3/4"	80'	3/4"	
2	Pipe Sections A, B & C	135,000	129	80'	1"	80'	1"	
3	Pipe Sections A, B, C & D	285,000	271	80'	1-1/4"	80'	1-1/4"	

DRAWING #12
Gas Piping – Residential
Use "Longest Length Method"

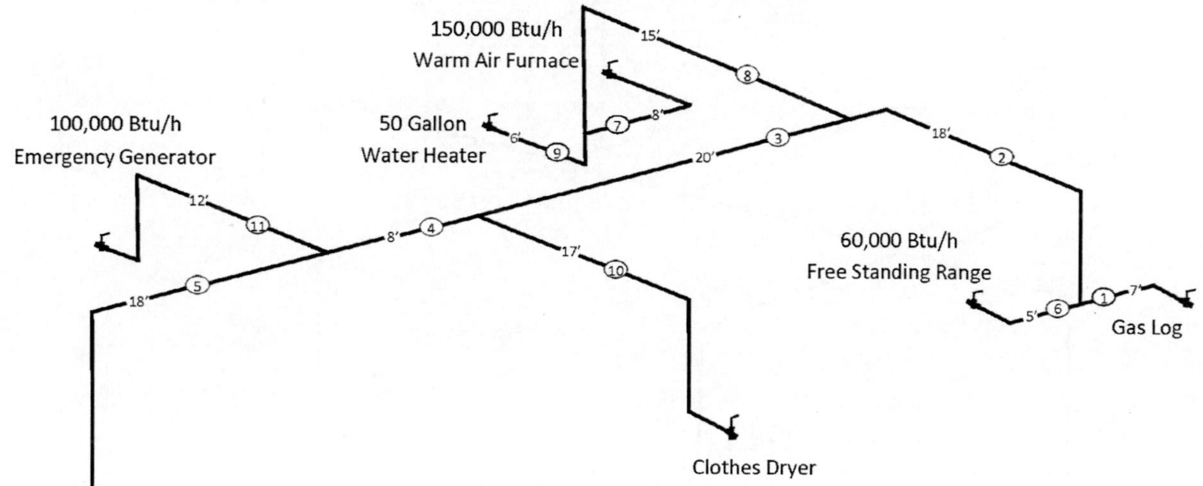

Given:

Gas Type: Natural

Pipe Material: Metallic Schedule 40

Inlet Pressure: Less than 2 psi

Pressure Drop: 0.5 in. w.c.

Heating Value: 1000 Btu/ft^3

Pipe Section	Appliance(s) Served	Btu/h	CFH Demand	Developed Length	Sizing Length	Sizing Table	Pipe Size
1							
2							
3							
4							
5							
6							
7							
8							
9							
10							
11							

DRAWING #13
Gas Piping – Residential
Use "Branch Length Method"

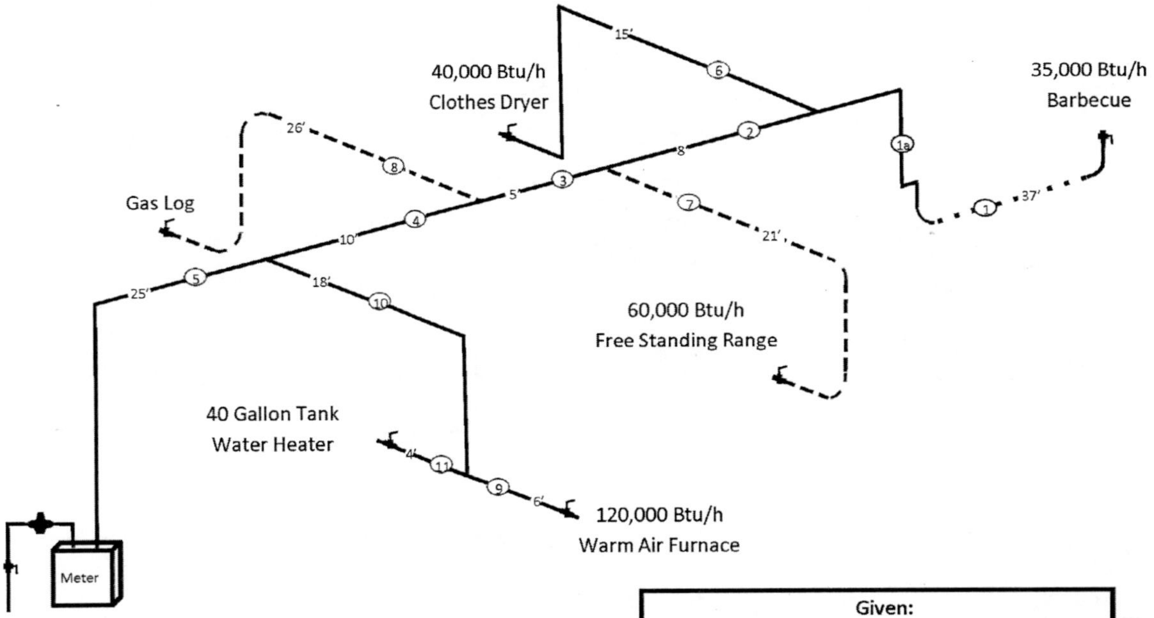

Given:

Gas Type: Natural

Pipe Material: Metallic Schedule 40 ———

CSST — — —

PE ··········

Inlet Pressure: Less than 2 psi

Pressure Drop: 0.5 in. w.c.

Heating Value: 900 Btu/ft^3

Pipe Section	Appliance(s) Served by Pipe Section	Btu/h	CFH Demand	Developed Length	Sizing Length	Sizing Table Used	Pipe Size
1							
1a							
2							
3							
4							
5							
6							
7							
8							
9							
10							
11							

DRAWING #14
Gas Piping – Residential

Use "Branch Length Method"

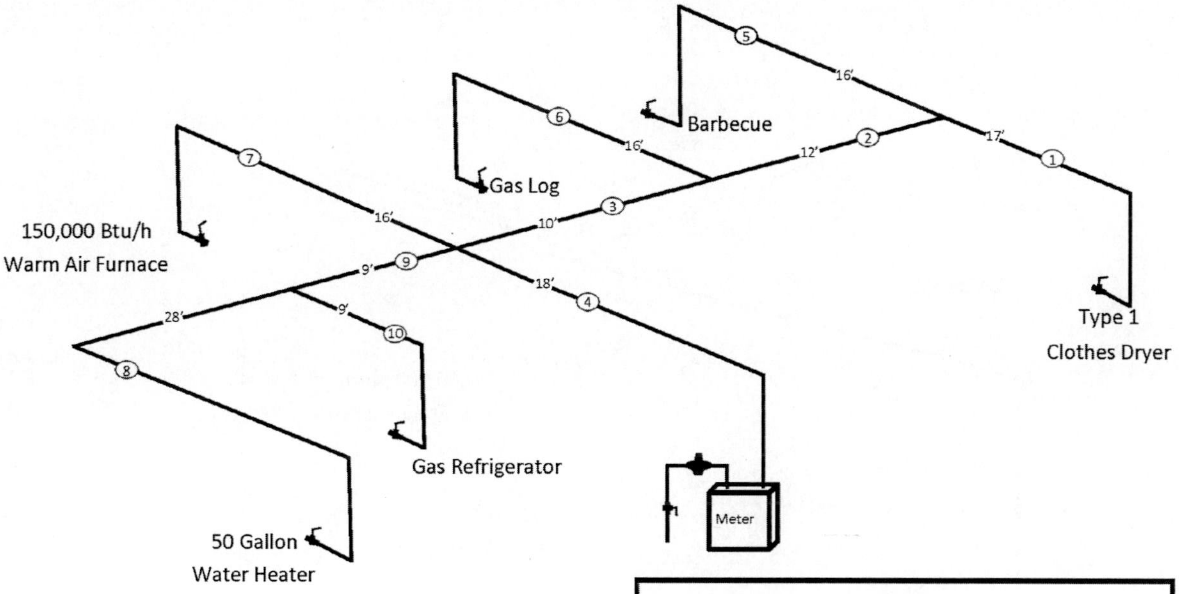

Given:

Gas Type: Natural

Pipe Material: Type L Copper Tubing

Inlet Pressure: Less than 2 psi

Pressure Drop: 0.5 in. w.c.

Heating Value: 1100 Btu/ft^3

Pipe Section	Appliance(s) Served	Btu/h	CFH Demand	Developed Length	Sizing Length	Sizing Table	Pipe Size
1							
2							
3							
4							
5							
6							
7							
8							
9							
10							

DRAWING #15
Gas Piping – Residential Building
Use "Longest Length Method"

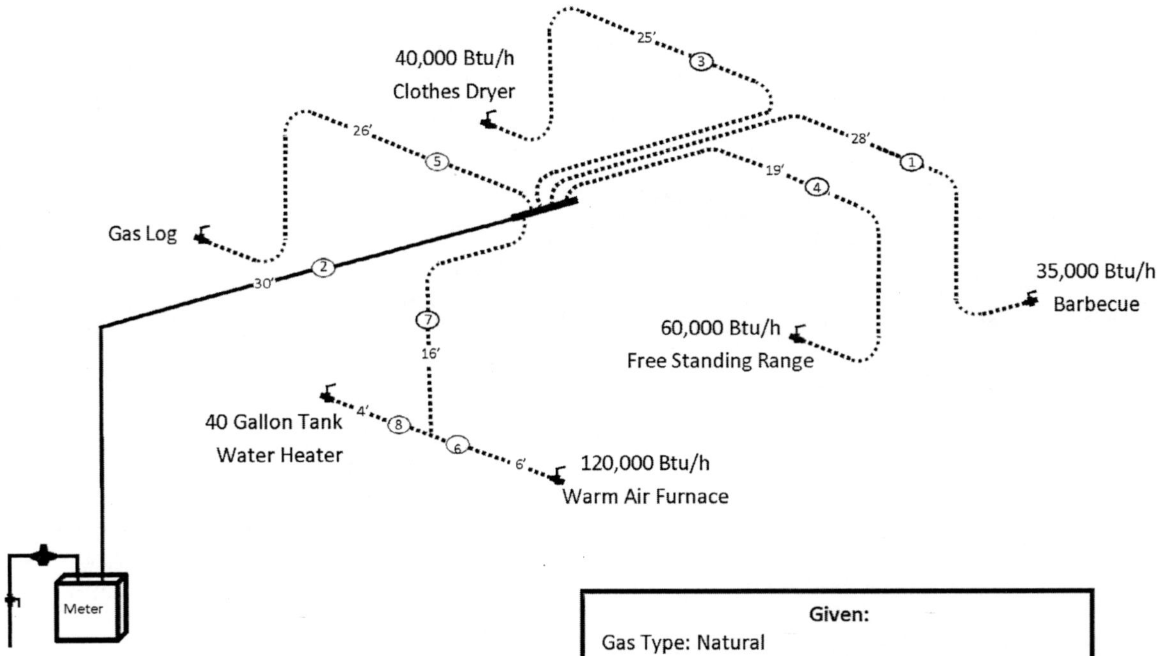

Given:

Gas Type: Natural

Pipe Material: Metallic Schedule 40 ———
CSST – – –

Inlet Pressure: Less than 2 psi

Pressure Drop: 0.5 in. w.c.

Heating Value: 900 Btu/ft^3

Pipe Section	Appliance(s) Served by Pipe Section	Btu/h	CFH Demand	Developed Length	Sizing Length	Sizing Table Used	Pipe Size
1							
2							
3							
4							
5							
6							
7							
8							

DRAWING #16
Gas Piping – Commercial Building

Use "Hybrid Method"

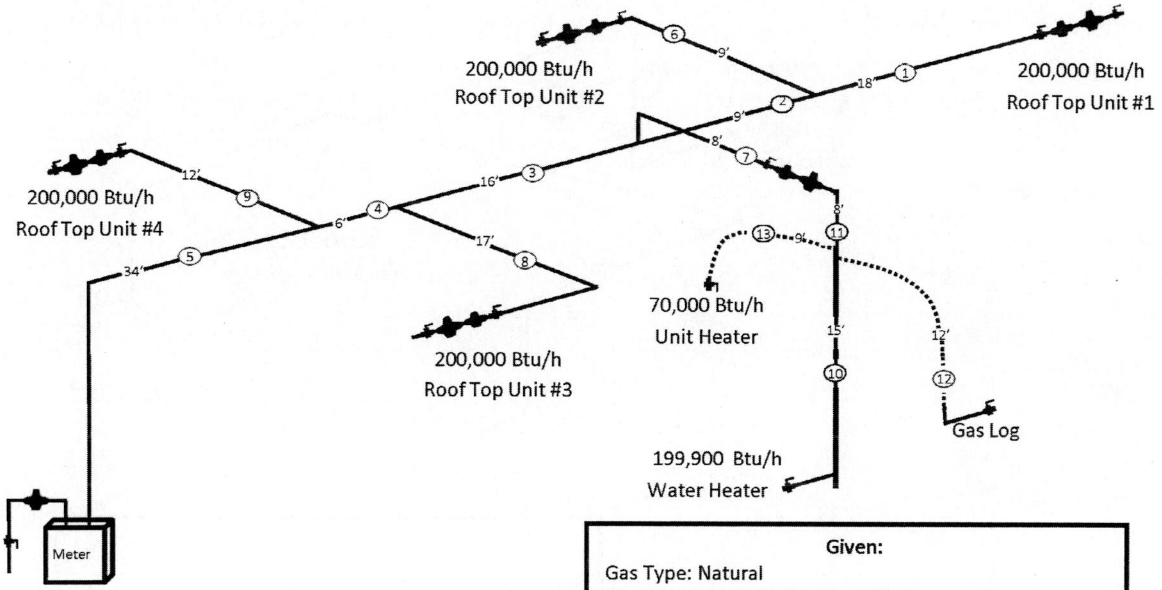

Given:
Gas Type: Natural
Pipe Material: Metallic Schedule 40 ———
CSST – – –
After meter regulator: Inlet Pressure: 2.0 psi
Pressure Drop: 1.0 psi
After line regulator: Inlet Pressure: less than 2.0 psi
Pressure Drop: 0.5 in. w.c.
Heating Value: 1035 Btu/ft^3
Pressure Regulator/Overpressure Protection

Pipe Section	Appliance(s) Served	Btu/h	CFH Demand	Developed Length	Sizing Length	Sizing Table	Pipe Size
1							
2							
3							
4							
5							
6							
7							
8							
9							
10							
11							
12							
13							

DRAWING #17
Gas Piping – Residential Building
Use "Hybrid Method"

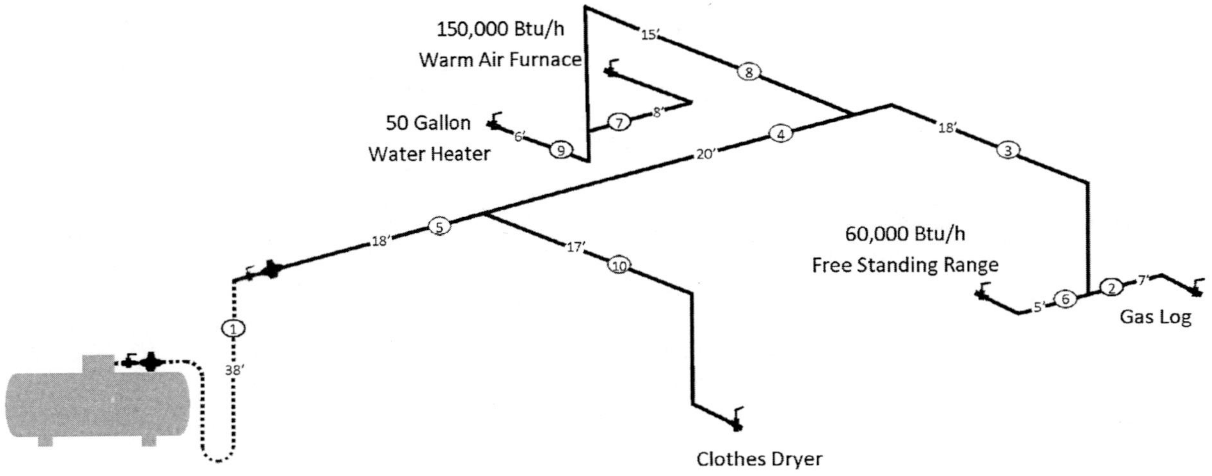

Given:

Gas Type: Undiluted Propane
Pipe Material: Metallic Schedule 40 ———
Type K Copper Tubing ··········
Downstream of Tank Regulator:
Inlet Pressure: 2.0 psi
Pressure Drop: 1.0 psi
Downstream of Line Regulator:
Inlet Pressure: 11.0 in. w.c.
Pressure Drop: 0.5 in. w.c.

Pipe Section	Appliance(s) Served	Btu/h	Developed Length	Sizing Length	Sizing Table	Pipe Size
1						
2						
3						
4						
5						
6						
7						
8						
9						
10						

PLUMBING FITTINGS EXAMINATION

The Uniform Plumbing Code requires that changes in direction of piping in plumbing systems be made by the use of approved fittings. With the exception of copper and plastic tubing for water systems, this would apply to all systems and especially to drainage and venting systems.

There are many fittings that have become standard use fittings and are in daily use. The fittings in this examination are basic to the trade. You should study these fittings and be able to recognize them so that you will know what they are and where they can be installed.

Included in this examination are the common type of hangers and supports used in plumbing systems.

A well-designed plumbing system includes the use of appropriate fittings and proper hangers and supports. This examination will help you obtain a good understanding of both.

Please note that some of the fittings depicted in this examination are no longer allowed in some jurisdictions.

PLUMBING FITTINGS EXAMINATION

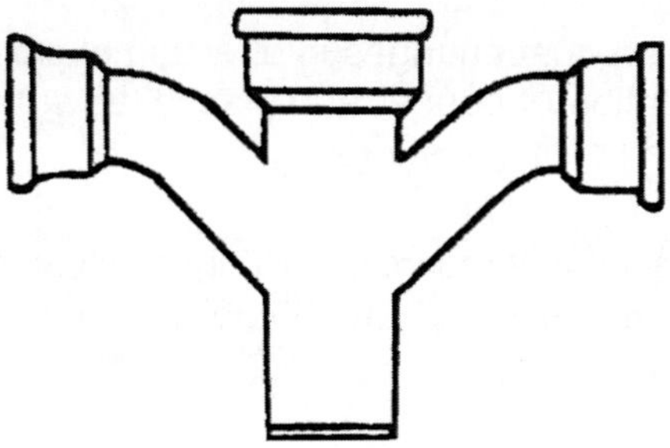

1. (A) double wye
   (B) double combination
   (C) double test tee
   (D) double sanitary tee

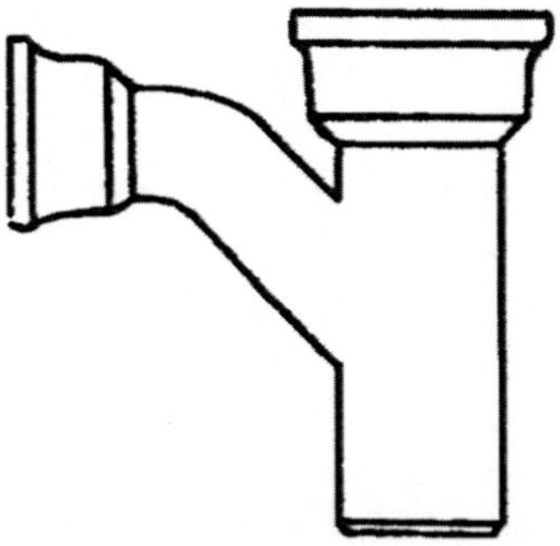

2. (A) soil 1/4 bend
   (B) sanitary tee
   (C) combination wye and 1/8 bend
   (D) soil wye

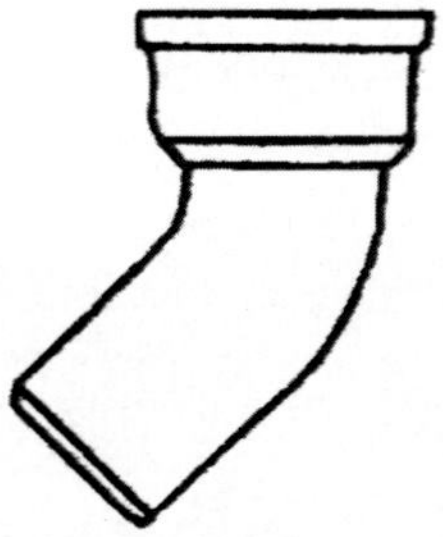

3. (A) soil 1/4 bend
   (B) medium sweep
   (C) double 1/4 bend
   (D) soil 1/8 bend

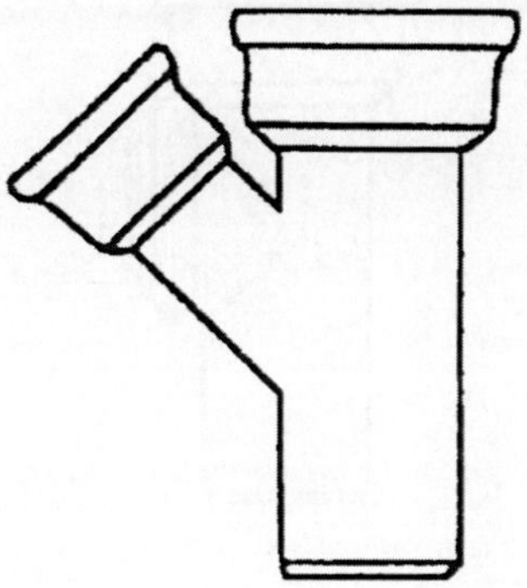

4. (A) upright wye
   (B) inverted wye
   (C) wye
   (D) tee wye

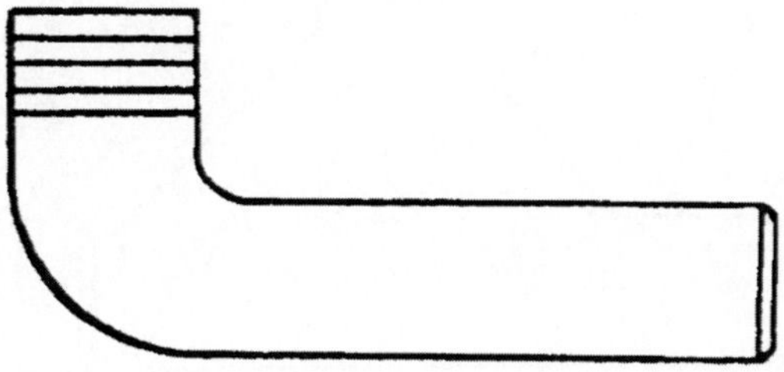

5. (A) regular closet bend
   (B) closet reducer
   (C) Arizona closet bend
   (D) extended closet bend

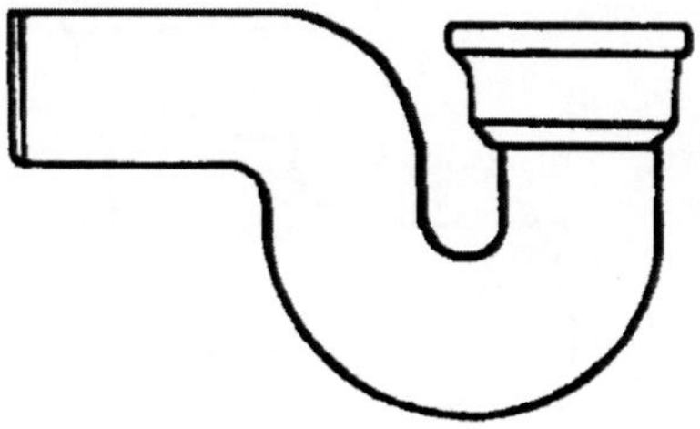

6. (A) running trap
   (B) soil S trap
   (C) soil P trap
   (D) drum trap

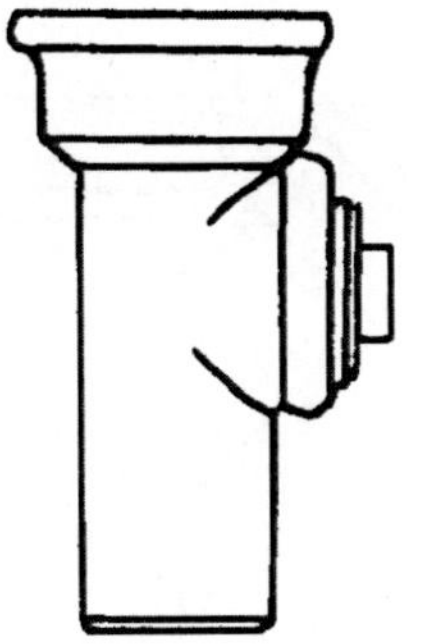

7. (A) sanitary tee
   (B) test tee
   (C) partition tee
   (D) dixie fitting

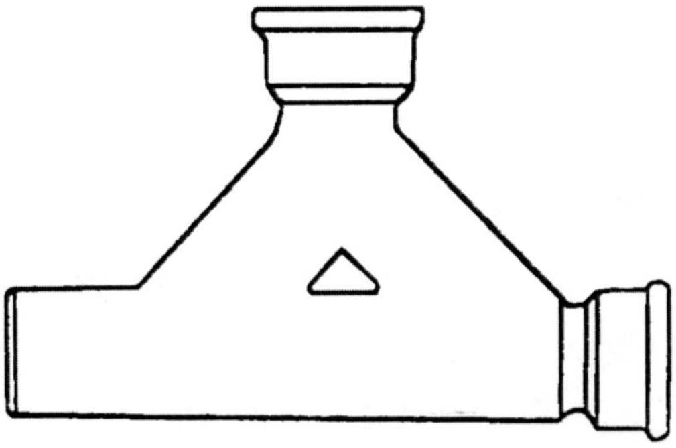

8. (A) figure #5 fitting
   (B) double half wye
   (C) double wye
   (D) two-way cleanout

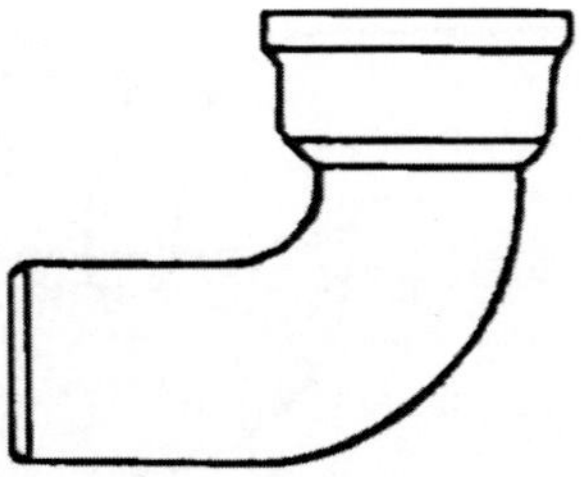

9. (A) soil 1/8 bend
   (B) soil 1/6 bend
   (C) soil 1/4 bend
   (D) long sweep bend

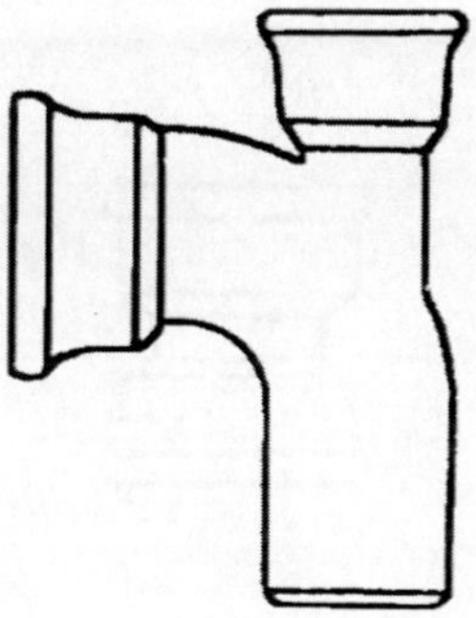

10. (A) side inlet 1/4 bend
    (B) double inlet 1/4 bend
    (C) low hub 1/4 bend
    (D) heel inlet 1/4 bend

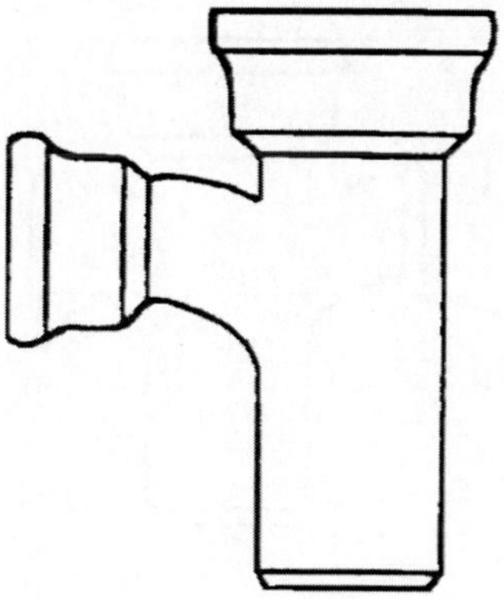

11. (A) straight tee
    (B) sanitary tee
    (C) combination wye and 1/8
    (D) tapped sanitary tee

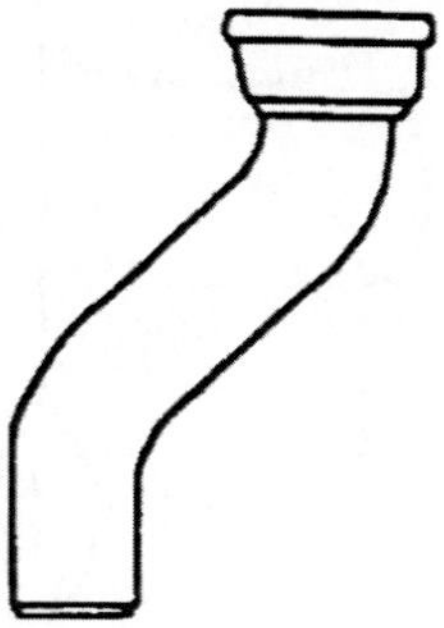

12. (A) soil offset
    (B) long 1/8 bend
    (C) sission joint
    (D) long 1/4 bend

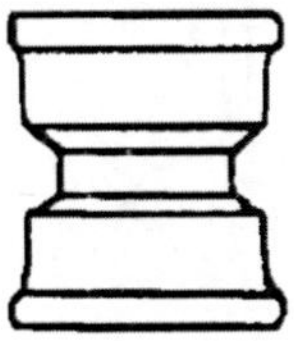

13. (A) double hub
    (B) soil sleeve
    (C) soil plug
    (D) soil reducer

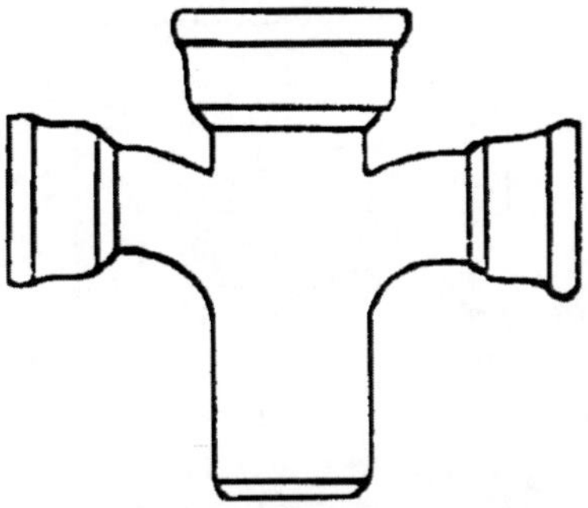

14. (A) double combination
    (B) double sanitary tee
    (C) straight cross
    (D) triple hub tee

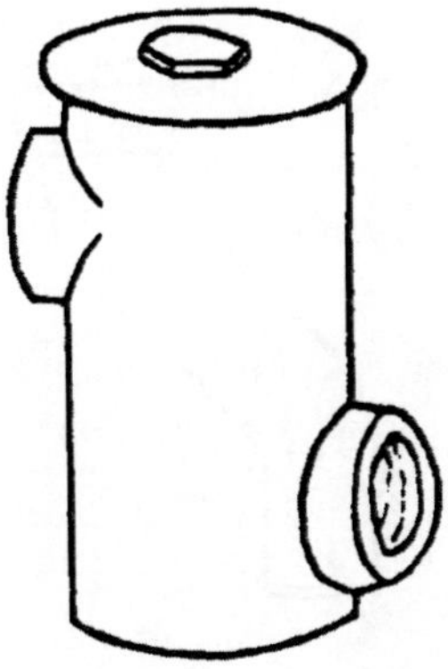

15. (A) bell trap
    (B) crown vented trap
    (C) partition trap
    (D) drum trap

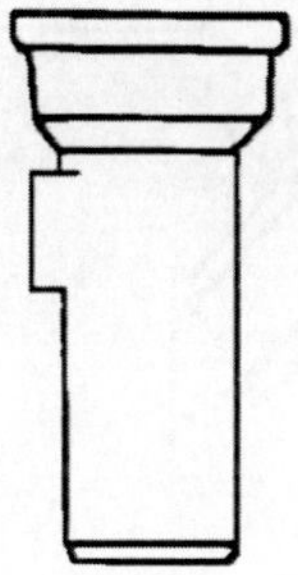

16. (A) sanitary tee
    (B) tapped straight tee
    (C) soil straight tee
    (D) fixture tee

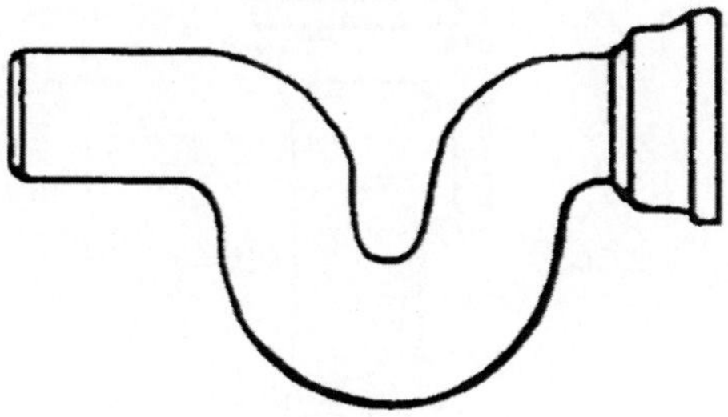

17. (A) P trap
    (B) S trap
    (C) crown vented trap
    (D) running trap

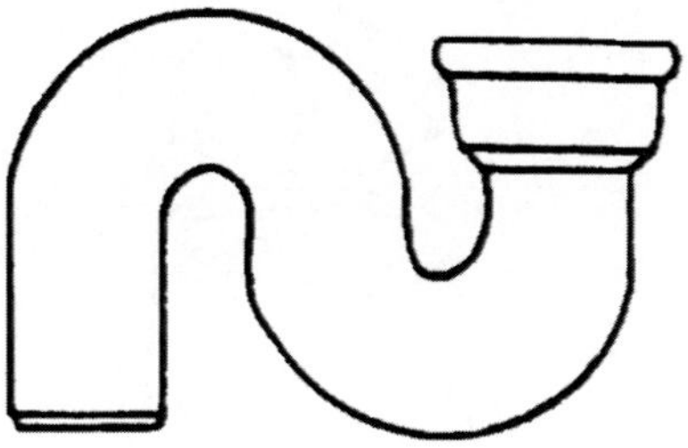

18. (A) P trap
    (B) S trap
    (C) crown vented trap
    (D) running trap

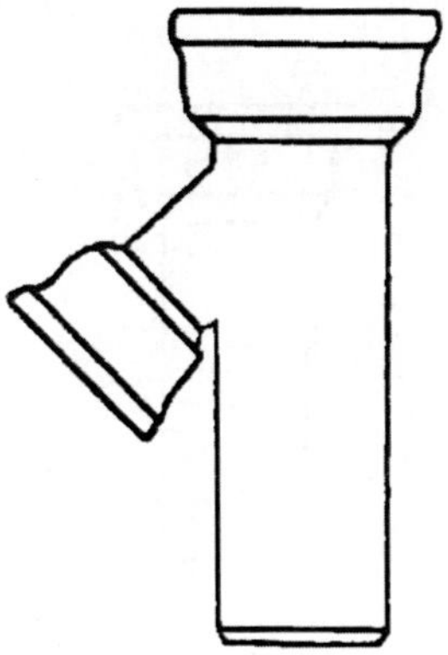

19. (A) inverted wye
    (B) upright wye
    (C) H branch
    (D) T branch

20. (A) sanitary tapped tee
    (B) tapped straight tee
    (C) soil straight tee
    (D) cross tee

21. (A) tucker
    (B) carlson fitting
    (C) soil sleeve with plug
    (D) iron body cleanout and plug

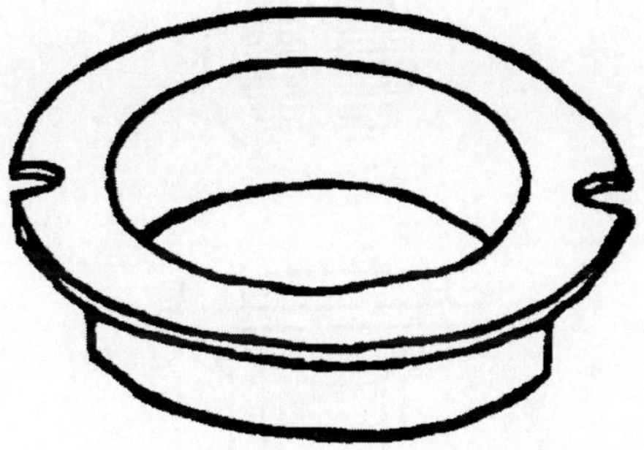

22. (A) offset closet ring
    (B) fixture
    (C) regular closet ring
    (D) closet bend support

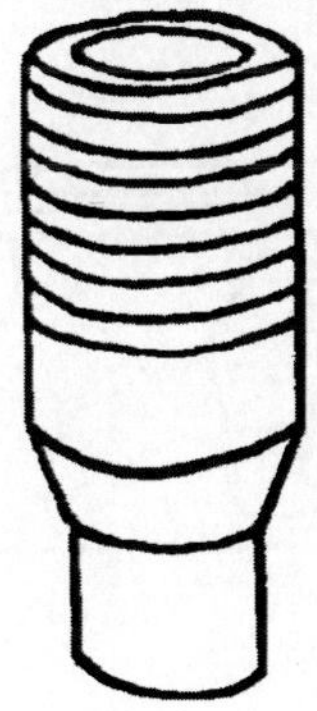

23. (A) soil sleeve
    (B) soil reducer
    (C) sission joint
    (D) closet reducer

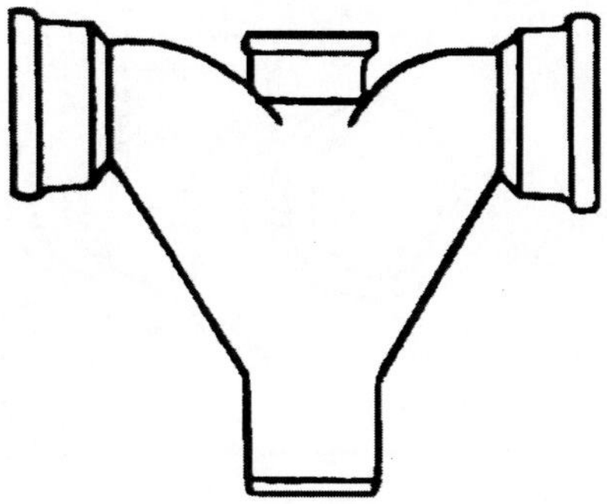

24. (A) double wye
    (B) figure #5
    (C) double sanitary tee
    (D) figure #3

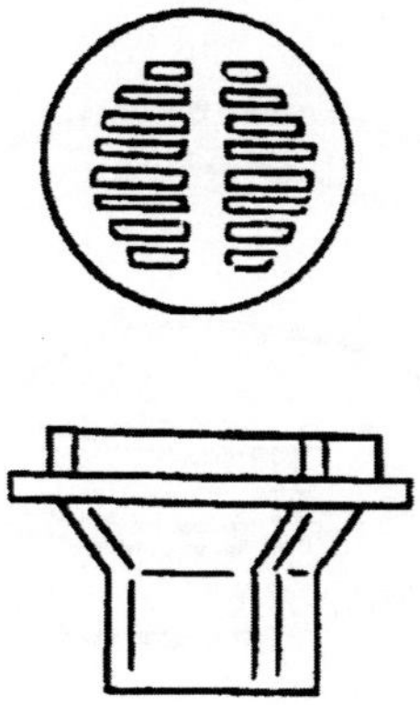

25. (A) frame and grate
    (B) catch basin
    (C) sand trap
    (D) floor drain

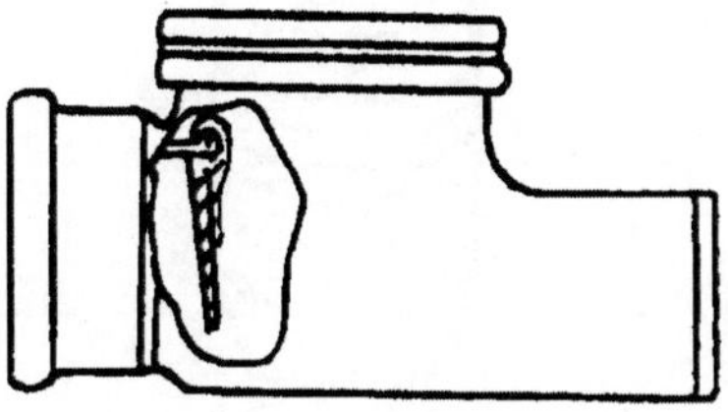

26. (A) house trap
    (B) flow control valve
    (C) backwater valve
    (D) ball check valve

27. (A) figure #6 single
    (B) figure #6 double
    (C) figure #8 single
    (D) figure #8 double

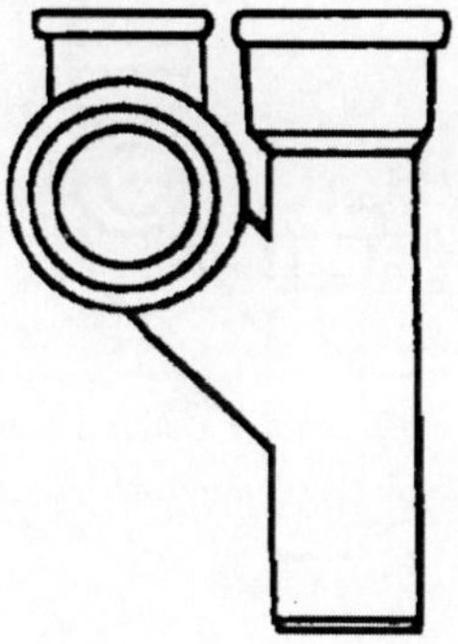

28. (A) figure #6 single
    (B) figure #6 double
    (C) figure #8 single
    (D) figure #8 double

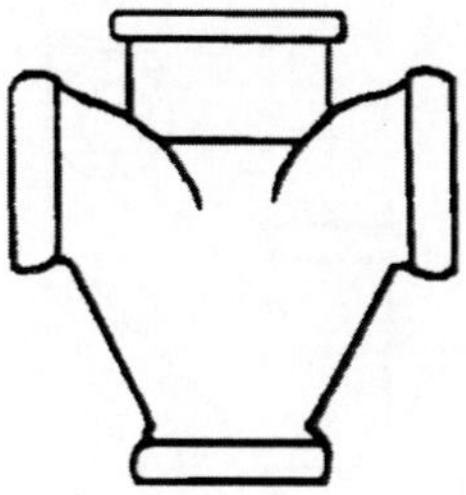

29. (A) figure #1
    (B) figure #2
    (C) figure #3
    (D) figure #4

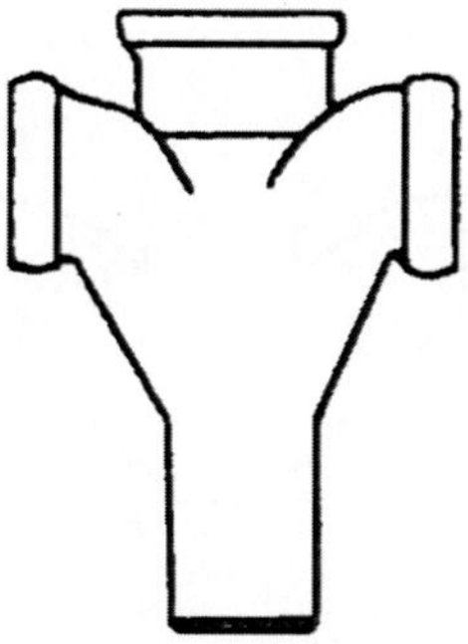

30. (A) figure #1
    (B) figure #2
    (C) figure #3
    (D) figure #4

31. (A) figure #6 single
    (B) figure #6 double
    (C) figure #4 single
    (D) figure #4 double

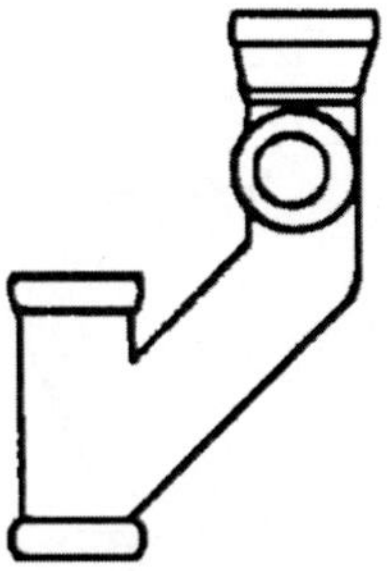

32. (A) figure #2 single
    (B) figure #2 double
    (C) figure #6 single
    (D) figure #6 double

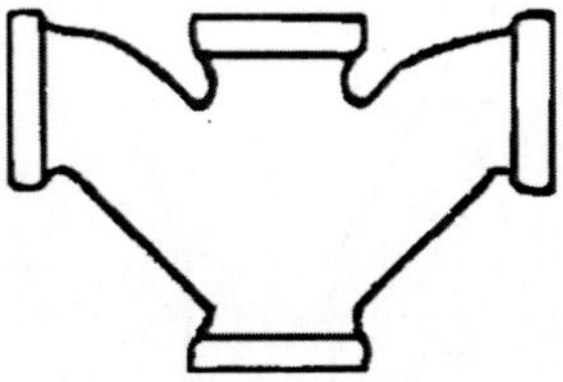

33. (A) double Durham wye
    (B) hub top cross tee
    (C) double Durham 90° long turn wye
    (D) double basin tee

34. 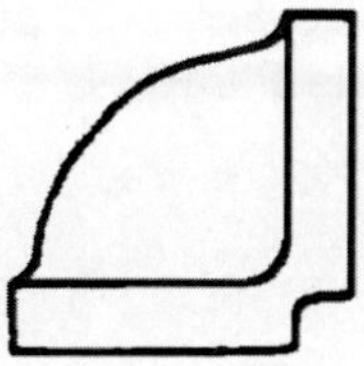

(A) Durham 22 1/2 degree ell
(B) Durham 60-degree ell
(C) Durham 60-degree ell
(D) Durham 90-degree ell

35. 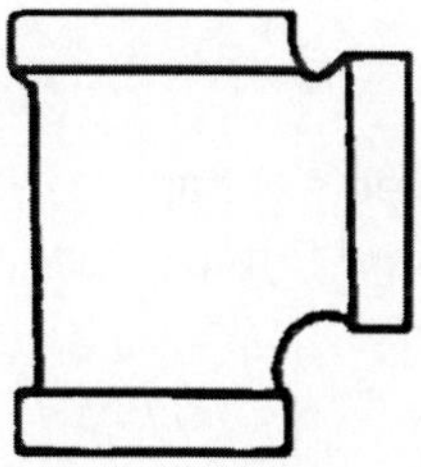

(A) Durham wye
(B) cast-iron tee
(C) Durham short-turn tee wye
(D) Durham long-turn tee

36. 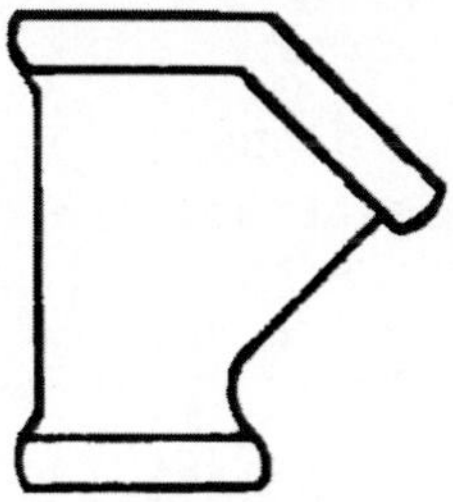

(A) Durham wye
(B) Durham tee
(C) Durham short-turn
(D) Durham long-turn tee wye

37. 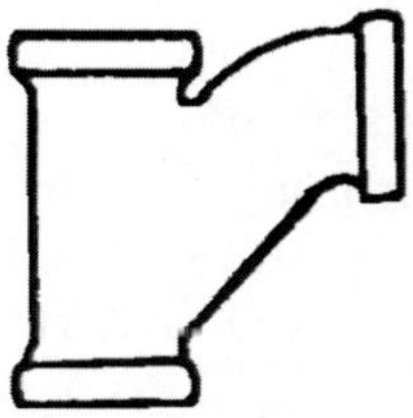

(A) Durham wye
(B) Durham tee
(C) Durham short-turn tee wye
(D) Durham long-turn tee wye

38. 

(A) Durham 45-degree ell
(B) Durham 60-degree ell
(C) Durham 72-degree ell
(D) Durham long 90 degree-ell

39. 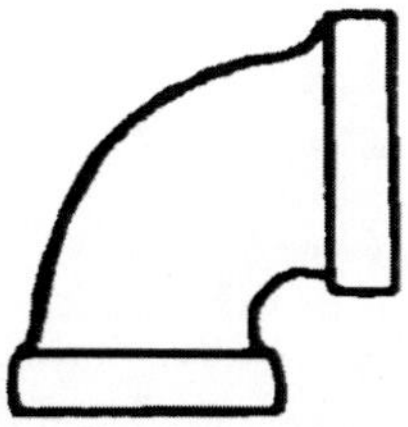

(A) Durham 45-degree ell
(B) Durham 60-degree ell
(C) Durham short 90-degree
(D) Durham long 90-degree

40.

(A) running trap
(B) deep seal trap
(C) adjustable P trap
(D) full S trap

41.

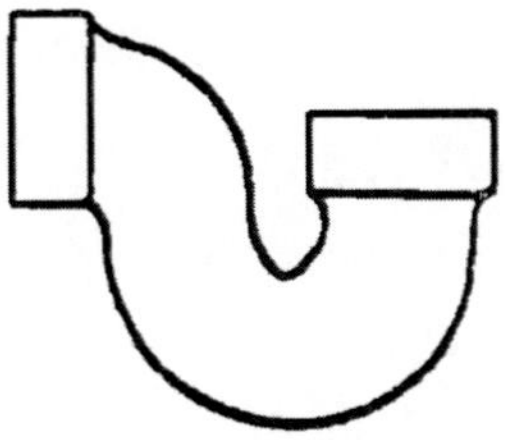

(A) running trap
(B) deep seal trap
(C) adjustable P trap
(D) standard P trap

42.

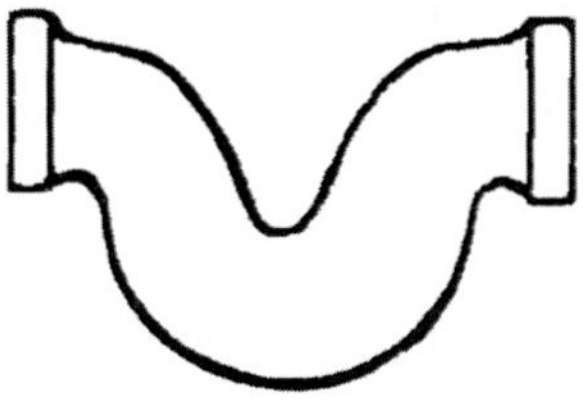

(A) running trap
(B) full S trap
(C) standard P trap
(D) deep seal trap

43.

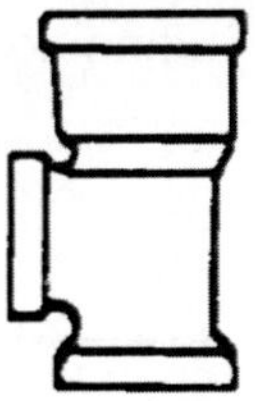

(A) hub top cross
(B) vent branch tee
(C) basin-sink tee
(D) tucker fitting

44.

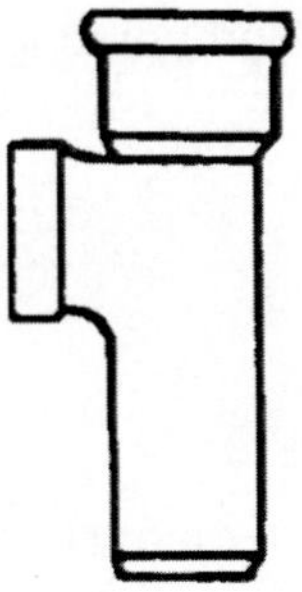

(A) straight tee
(B) branch tee
(C) waste union
(D) single hub tap tee

45. 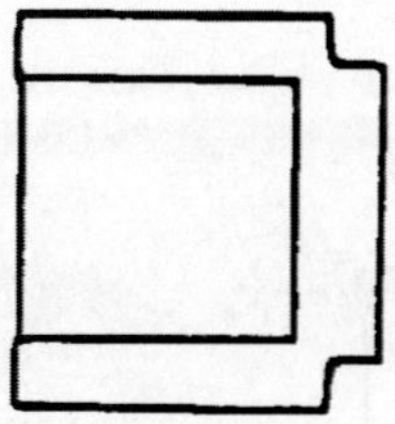

(A) regular vent tee
(B) cross vent tee
(C) reducing vent tee
(D) reducing vent cross

46. 

(A) regular vent tee
(B) tapped vent tee
(C) reducing vent tee
(D) tapped vent cross

47. 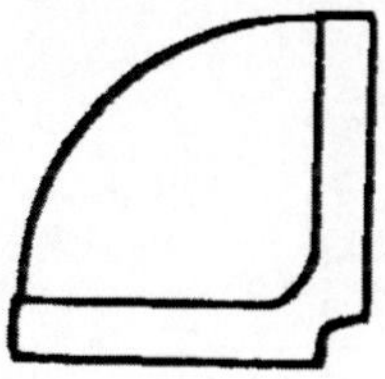

(A) 90-degree vent ell
(B) 72-degree vent ell
(C) 45-degree vent ell
(D) 60-degree vent ell

48. 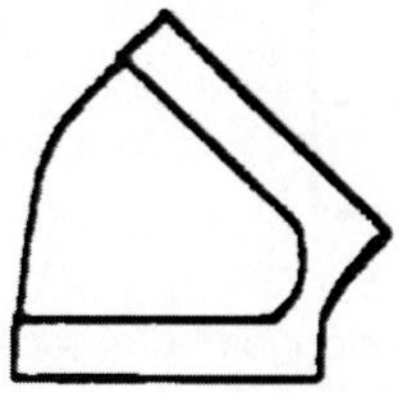

(A) 90-degree vent ell
(B) 72-degree vent ell
(C) 45-degree vent ell
(D) 60-degree vent ell

49. 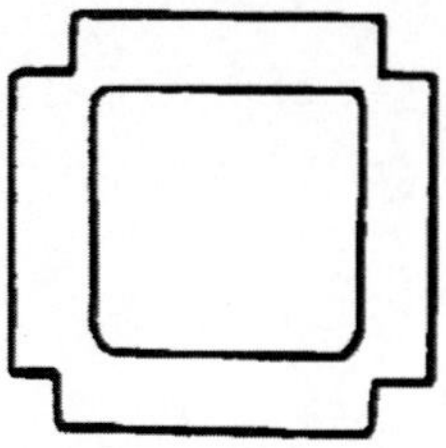

(A) regular vent tee
(B) cross vent tee
(C) reducing vent tee
(D) reducing vent cross

50. 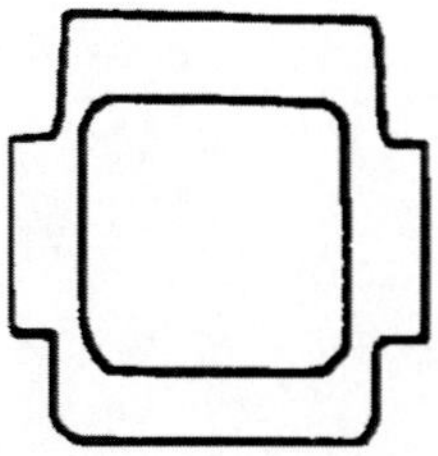

(A) regular vent tee
(B) cross vent tee
(C) reducing vent tee
(D) reducing vent cross

51. (A) cast-iron hubless drainage fittings
    (B) cast-iron Durham drainage fittings
    (C) copper solder joint drainage fittings
    (D) plastic ABS drainage fittings

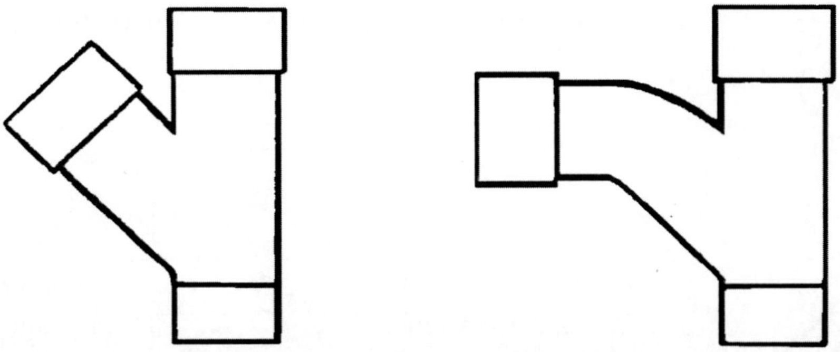

52. (A) cast-iron hubless drainage fittings
    (B) cast-iron Durham drainage fittings
    (C) Dual-Tite compression joint drainage fittings
    (D) plastic ABS or PVC drainage fittings

53.

(A) riser clamp
(B) clevis hanger
(C) pipe saddle
(D) one hole clamp

54.

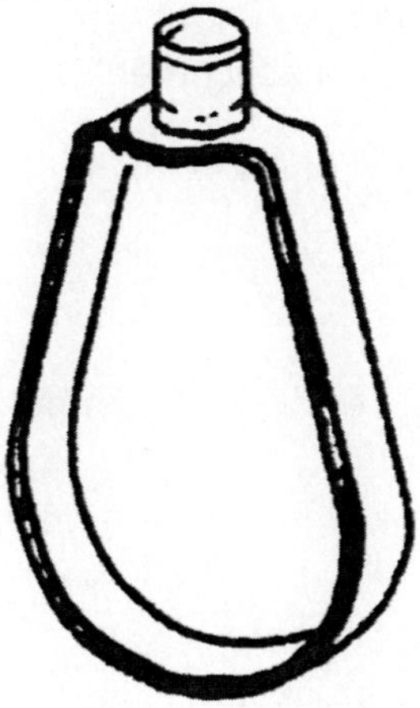

(A) clevis hanger
(B) E-Z grip hanger
(C) pipe saddle
(D) band hanger

55.

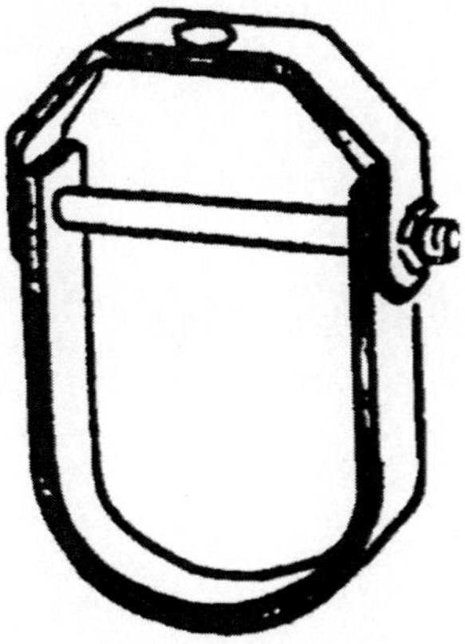

(A) clevis hanger
(B) E-Z grip hanger
(C) pipe saddle
(D) band hanger

56.

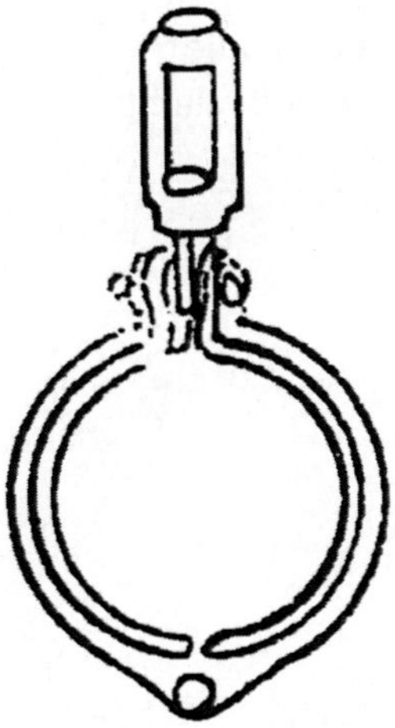

(A) clevis hanger
(B) straight hook
(C) ring and bolt hanger
(D) split ring hanger

57.

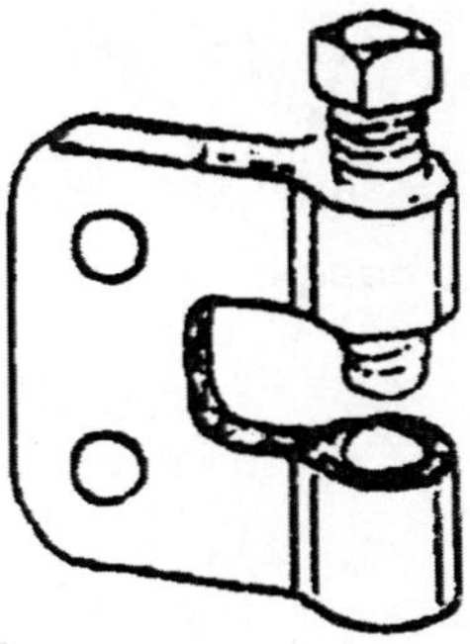

(A) conduit clamp
(B) angle bracket
(C) "C" clamp
(D) rod beam attachment

58.

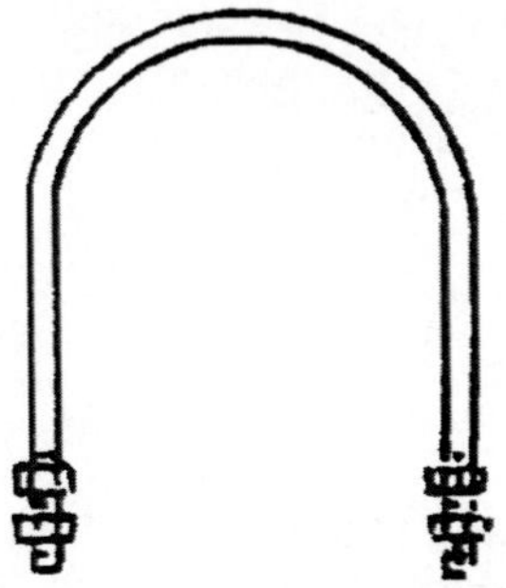

(A) one hole clamp
(B) "U" bolt
(C) roller hanger
(D) angle bracket

59.

(A) one hole clamp
(B) "U" bolt
(C) side beam clamp
(D) pipe clamp

60. 

(A) clevis hanger
(B) straight hook
(C) roller hanger
(D) pipe clamp

61 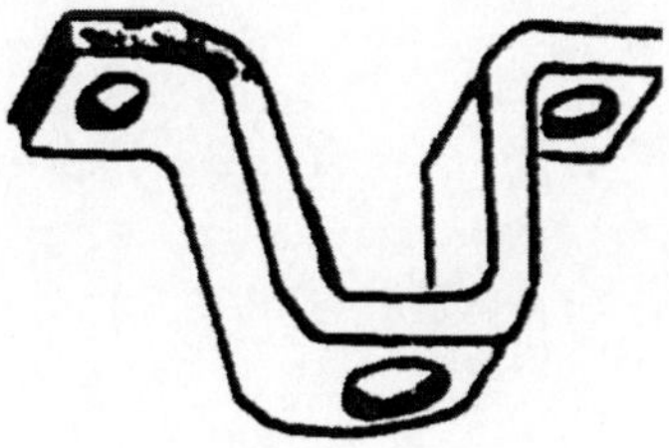

(A) conduit clamp
(B) "U" hanger
(C) angle bracket
(D) rod beam attachment

62. 

(A) hanger rod clip
(B) one hole clamp
(C) straight hook
(D) pipe clamp

63. 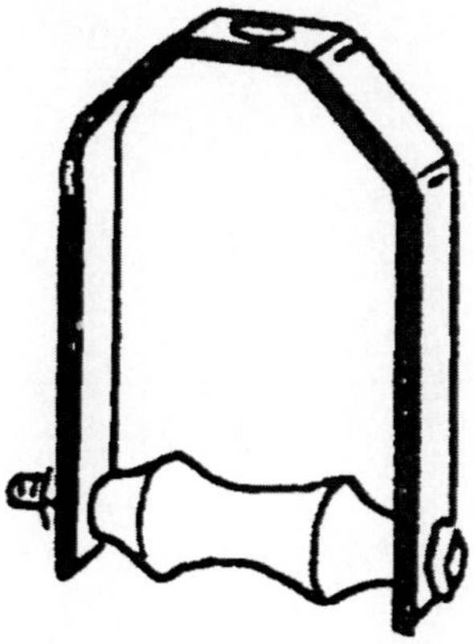

(A) clevis hanger
(B) "U" hanger
(C) roller hanger
(D) hanger rod clip

ANSWERS

The correct section number is noted in italics.

CHAPTER 1 – ADMINISTRATION

1 A *101.1*
2 C *103.3*
3 A *101.2*
4 C *102.4.2*
5 C *102.3*
6 D *102.2*
7 B *102.1*
8 B *102.6*
9 C *102.7*
10 A *103.1*
11 C *106.4*
12 C *104.1*
13 C *104.3*
14 B *105.1*
15 C *105.3*
16 A *105.2.3*
17 B *105.2.4*
18 D *105.2.5*
19 B *105.4*
20 D *101.3*
21 B *101.4*
22 A *102.8*
23 B *105.4*

CHAPTER 2 – DEFINITIONS

1 B *204.0*
2 A *203.0*
3 A *204.0*
4 C *205.0*
5 C *205.0*
6 D *206.0*
7 C *208.0*
8 D *224.0 and 217.0*
9 A *218.0*
10 C *221.0*
11 D *225.0*
12 A *224.0*
13 A *222.0*
14 D *222.0*
15 D *218.0*
16 B *227.0*
17 D *221.0*
18 C *211.0*
19 B *209.0*
20 B *205.0*
21 A *203.0*
22 C *210.0*
23 B *225.0*
24 A *222.0*
25 A *218.0*
26 B *225.0*
27 C *204.0*
28 B *218.0*
29 C *211.0*
30 A *215.0*
31 A *222.0*
32 A *218.0*
33 A *221.0*
34 D *205.0*
35 D *215.0*
36 A *212.0*
37 D *203.0*
38 B *204.0*
39 A *205.0*
40 D *204.0*
41 D *225.0*
42 C *204.0*
43 A *224.0*
44 D *221.0*
45 B *205.0*
46 B *205.0*
47 A *225.0*
48 B *224.0*
49 A *206.0*
50 A *224.0*
51 B *221.0*
52 A *209.0*
53 C *221.0*
54 A *224.0*
55 B *204.0*
56 B *224.0*
57 C *222.0*
58 D *209.0*
59 B *220.0*
60 C *208.0*
61 C *209.0*
62 A *203.0*
63 A *224.0*
64 D *221.0*
65 B *205.0*
66 A *215.0*

CHAPTER 3 – GENERAL REGULATIONS

1 C *303.1 and 304.1*
2 A *303.1*
3 C *309.3*
4 B *307.1*
5 C *308.1*
6 A *309.1*
7 B *Table 313.6*
8 B *311.1 Exception*
9 A *314.1*
10 A *Table 313.3*
11 B *Table 313.3*
12 A *313.5*
13 B *Table 313.3*
14 A *Table 313.3*
15 B *Table 313.3*
16 B *314.4*
17 C *Table 313.3*
18 D *Table 313.3 Note 2*
19 B *313.1*
20 C *312.1*
21 A *301.2*
22 C *301.3*
23 B *302.1*
24 A *312.2*
25 B *312.12.3*
26 B *310.2*
27 D *301.2*
28 C *312.10*
29 A *Table 313.3*
30 C *319.1*
31 B *317.1(4)*
32 B *318.5*
33 B *301.3*
34 C *301.4 Exception*
35 D *309.5*
36 C *Table 313.3*

CHAPTER 4 – PLUMBING FIXTURES AND FIXTURE FITTINGS

1 A *402.3*
2 A *402.10*
3 A *404.2*
4 C *409.2*
5 D *419.2*
6 A *419.2*
7 D *405.3*
8 B *402.5*
9 C *402.5*
10 A *413.2*
11 B *412.2*
12 C *405.2*
13 B *408.7(2)*
14 D *408.6*
15 A *408.6*
16 C *408.8*
17 D *408.8.4*
18 A *415.4*
19 D *402.5*
20 B *402.5*
21 A *413.2*
22 D *402.2*
23 A *406.2*
24 C *408.9*
25 A *402.4*
26 A *407.5*
27 B *408.8*
28 D *408.5 and 418.2*
29 D *408.4*
30 C *402.5*
31 D *412.1.1*
32 B *410.3*
33 B *402.6.1*
34 C *402.6*
35 C *402.6*
36 B *409.5*
37 D *414.3*
38 A *416.4*
39 A *402.4*
40 D *416.2*
41 C *Table 422.1*
42 A *422.2(4)*

CHAPTER 5 – WATER HEATERS

1 B *502.1*
2 B *509.3.3.3 and 509.10.3.2*
3 A *509.10.12*
4 C *509.10.1.1*
5 D *501.1 and Table 501.1(2)*
6 B *507.13.1*
7 C *503.2*
8 C *504.2*
9 C *Table 501.1(2)*
10 C *506.4.1 and 506.4.2*
11 B *506.2.1*
12 A *506.9(3)*
13 C *506.9(5)*
14 C *504.1(2)*
15 B *507.4*
16 D *507.5*
17 B *509.4.1*
18 C *509.4.3*
19 B *509.8.3*
20 D *509.12.4*
21 C *510.2.1 and Table 510.2.1*
22 C *509.8.1 and Figure 509.8.1*
23 B *509.8.1 and Figure 509.8.1*
24 B *509.6.1(1)(b)*
25 D *508.3.1*
26 A *509.10.2.3*
27 A *507.2*
28 B *506.8*
29 A *509.13*
30 C *504.3.1*
31 C *507.13*
32 A *507.24*
33 C *509.6.1(2)*
34 C *507.13.2*
35 D *505.4.1*
36 B *504.3.1*
37 C *508.4.2*
38 D *507.26*
39 A *509.3.6*
40 D *509.4.1*
41 A *509.6*
42 B *510.2.6*

CHAPTER 6 – WATER SUPPLY AND DISTRIBUTION

1 A *601.2*
2 C *602.3*
3 B *603.5.2 and Table 603.2*
4 C *603.5.1*
5 C *603.5.5(2)*
6 B *Table 603.2*
7 B *604.13*
8 D *601.3.5*
9 D *Table 604.1*
10 B *604.7*
11 D *606.1*
12 A *609.1*
13 B *610.7(5)*
14 B *604.3 Exception*
15 A *606.5*
16 C *608.1*
17 C *608.2*
18 D *608.6*
19 A *608.3*
20 A *609.3(1)*
21 D *604.6*
22 A *609.2 and Table 701.2*
23 A *610.8(2)*
24 B *610.8(6) and Table 610.4 Note 2*
25 C *609.3(2)*
26 B *Table 604.1*
27 C *604.13*
28 D *609.5*
29 A *606.2*
30 C *603.2*
31 B *610.12.1*
32 A *608.5*
33 C *610.3*
34 B *610.13(6)*
35 B *604.4*
36 A *604.4*
37 B *605.1.2*
38 B *609.5*
39 A *608.1*
40 B *609.4 Exception*
41 C *609.4*
42 C *601.2.1*
43 D *601.3.2*
44 A *604.10.1*
45 B *608.3*
46 A *605.5.2*
47 A *605.1.1 and 605.1.4*
48 A *605.1.4*
49 B *605.2.3 and 605.12.3*
50 C *605.1.1*
51 B *605.1.3.3*
52 A *605.6.1 and 605.11.1*
53 D 601.3.3(1), (2) and (4)
54 D *612.3.5*
55 A *612.4.5*
56 A *612.4.2 1*
57 B *Table 612.3.3.1*
58 B *Table 612.3.3.1 Note 1*
59 C *612.5.3.1*
60 C *612.3.3.1(2)*
61 A *612.5.3; 612.5.3.2.2, Table 612.5.3.2(1) and Table 612.5.3.2(2)*
62 A *612.7.1(8) to 609.4*
63 C *612.4.2; 612.7.1(6)*
64 B *612.5.2.2(1)*
65 B *603.5.22*
66 D *605.2.1*
67 B *605.2.1.1 and 605.9.3*
68 C *Table 610.3*
69 D *609.4*
70 A *608.3*

CHAPTER 7 – SANITARY DRAINAGE

1 D *Table 703.2*
2 A *708.1*
3 D *Table 702.1(1)*
4 D *703.1*
5 A *Table 702.1*
6 A *Table 702.2(2)*
7 A *Table 702.1 Footnote #2*
8 C *706.3*
9 C *707.9*
10 A *710.4*
11 B *710.3(2)*
12 D *710.5*
13 C *Table 703.2 Footnote #5*
14 A *Table 702.1*
15 B *Table 703.2*
16 A *Table 703.2*
17 B *Table 703.2*
18 B *Table 702.1*
19 D *706.4*
20 B *707.10 and Table 707.1*
21 C *Table 702.1*
22 D *711.1*
23 D *711.1*
24 D *706.4*
25 D *Table 703.2 Footnote #5*
26 B *717.1*
27 C *719.1*
28 A *718.1 Exception*
29 D *719.6*
30 D *701.2(2)*
31 C *701.3.1*
32 A *712.1*
33 B *701.3.2*
34 B *Table 703.2 Footnote #5*
35 C *710.13*
36 C *710.13.1*
37 B *Table 702.1*
38 B *707.4*
39 B *722.1*
40 D *720.0*
41 B *707.3*
42 A *Table 702.1*
43 A *705.2.1*
44 B *713.1*
45 B *705.5*
46 B *715.3.1*
47 B *710.9*

CHAPTER 8 – INDIRECT WASTES

1 C *801.2*
2 B *801.3.1*
3 C *801.3.2*
4 A *801.5*
5 C *801.6*
6 B *801.7*
7 C *802.1*
8 B *801.4*
9 D *804.1*
10 B *804.1*
11 A *804.1*
12 B *805.1*
13 C *807.3*
14 D *809.1*
15 C *810.1*
16 A *810.4*
17 D *811.1*
18 C *811.2*
19 B *811.4*
20 C *811.5*
21 C *811.6*
22 B *811.7*
23 C *812.1*
24 D *813.1*
25 A *814.1*
26 B *814.1*
27 C *814.1*
28 A *801.4*
29 A *814.3 and Table 814.3*
30 D *814.5*

CHAPTER 9 – VENTS

1 B *901.2*
2 C *902.1*
3 A *902.2*
4 B *904.1*
5 D *904.1*
6 B *904.2 and Table 703.2 Note 6*
7 A *904.1*
8 C *905.3*
9 A *905.2*
10 B *222.0, 905.5*
11 C *904.1 and Tables 702.1 and 703.2*
12 A *906.1*
13 B *906.1*
14 C *906.2*
15 C *906.2*
16 A *906.3*
17 D *906.5*
18 B *907.1*
19 C *907.1*
20 C *904.1, and Tables 702.1 and 703.2*
21 D *905.1*
22 C *904.1*
23 C *904.2*
24 D *903.3*
25 B *907.1 and 907.2*
26 D *907.2*
27 A 907.1
28 B *905.3*
29 A *906.7*
30 A *903.2.2*
31 D *903.1(2)*
32 B *906.6(2)*
33 B *908.1.1*
34 D *by definition*
35 C *Table 702.1*
36 B *Tables 702.1 and 703.2*
37 C *908.1 and Table 702.1*
38 C *909.1*
39 A *909.1*
40 B *909.1*
41 B *909.1 and Table 702.1 Note 2*
42 D *706.4*
43 A *909.1*
44 C *910.3*
45 B *on drawing and Appendix B - B101.6.1*
46 C *910.4*
47 B *910.3*
48 C *910.7 Note, Appendix B - B101.4*
49 C *910.3*
50 C *908.2*
51 C *908.2.1*
52 B *901.3*
53 B *908.2.2*
54 D *908.2.1*
55 C *908.2.4*
56 A *912.1*
57 C *911.1*
58 D *911.1*
59 C *911.3.2*
60 C *911.2*
61 D *911.4*
62 A *911.4.1.1*
63 B *911.3*
64 A *911.5*
65 B *906.1*
66 D *904.1 Exception*
67 D *204.0*
68 B *906.1*
69 A *910.4*

CHAPTER 10 – TRAPS AND INTERCEPTORS

1 A *1001.2*
2 C *1001.2*
3 B *1001.2*
4 D *1001.2*
5 B *1002.1*
6 C *1002.3*
7 B *1002.4*
8 B *1002.2 and Table 1002.2*
9 A *1003.1*
10 A *1003.2*
11 D *1003.3*
12 D *1004.1*
13 C *1005.1*
14 C *1006.1*
15 C *1007.1*
16 C *1008.1*
17 D *1009.1*
18 B *1009.3*
19 D *1009.4 and 1009.5*
20 D *1011.1*
21 C *1012.1*
22 C *1014.3.6, Table 1014.3.6 and Tables 703.2 and 702.2(2)*
23 B *1014.2*
24 B *1014.1.1*
25 C *1014.2.1*
26 C *1016.1 and 1016.2*
27 C *1016.3*
28 B *1016.4*
29 D *1017.1*
30 B *1017.4(2)*
31 C *1014.1.2*
32 B *1014.2.2*
33 B *1014.3.4*
34 C *1014.3.6 and Table 1014.3.6*
35 A *1014.3.4.1*
36 D *1014.3.5*
37 B *1014.3.7*
38 D *1014.1*
39 D *1003.2*
40 A *1014.3.4*
41 A *Tables 702.1, 702.2 and 1014.3.6*
42 B *Tables 702.1, 702.2 and 1014.2.1*
43 A *1014.3.4.2*

CHAPTER 11 – STORM DRAINAGE

1 B *1101.3*
2 D *1101.6.1*
3 C *1101.6*
4 B *1101.14*
5 C *1101.12.1*
6 B *1101.12.2.2*
7 A *1101.15*
8 B *1101.4.2*
9 C *1103.2 and Table 1103.2 Note 1*
10 A *1102.2*
11 D *1102.3*
12 A *1105.1(3)*
13 A *1107.2.1*
14 B *1107.2*
15 D *1105.1(13)*
16 C *1103.1 and Table 1103.1*
17 D *1103.2 and Table 1103.2*
18 D *1103.3*
19 A *1103.4(1)*
20 D *1101.6*
21 A *1101.16.1*
22 B *1101.12.2.2.1*
23 B *1101.4*
24 D *Table 1103.1 Note 3*
25 D *1102.2 Exception*
26 C *1101.15.2*
27 D *1103.2 and Table 1103.2*
28 D *1101.2*
29 C *1106.1*

CHAPTER 12 – FUEL GAS PIPING

1 C *1206.1*
2 B *1208.1*
3 A *1210.1.7*
4 A *1208.5.8.2 and Table 1208.5.8.2*
5 B *1208.5.9.1(2)*
6 A *1210.3.5.1 and Table 1210.3.5.1*
7 D *1208.5.12.4*
8 D *1208.5.11.7*
9 D *1210.3.3*
10 B *1210.1.1*
11 C *1210.1.6(1)*
12 A *1210.4.2*
13 A *1210.5.3*
14 A *Table1208.3.1*
15 C *1208.15(1)*
16 C *1208.5.3.4*
17 B *1210.6.1(5)*
18 B *1208.6.1.1*
19 B *1213.1.6*
20 C *1210.1.1*
21 A *1210.9*
22 D *1208.7.2*
23 C *1208.6.3*
24 B *1212.6*
25 C *1212.6*
26 C *1212.6 Exception 2*
27 C *1203.3.1*
28 D *1203.3.1*
29 B *1204.1*
30 B *1206.3*
31 B *1203.1*
32 A *1213.4.1*
33 D *1208.5.3.5*
34 D *1213.4.2*
35 C *1212.11*
36 B *1208.5.12.5*
37 A *1208.7.3*
38 B *1208.15(5)*
39 A *1214.4*
40 B *1214.4*
41 A *1208.5.2.1*
42 B *1214.2*
43 A *1214.3*
44 B *1208.5.2.4*
45 A *1210.1.7.2(2)*
46 C *1212.3.1(2)*
47 A *1208.5.5*
48 D *1209.1*
49 C *1210.9.4*
50 A *1211.2.2*
51 D *1213.3*
52 C *1208.8*
53 B *1208.10*
54 C *1208.10.1*
55 D *1210.1.3.2*
56 D *1210.1.3.4 and 1210.1.3.6*
57 D *1210.1.3.5*
58 D *1208.11.1*

CHAPTER 13 – HEALTH CARE FACILITIES AND MEDICAL GAS AND MEDICAL VACUUM SYSTEMS

1 D *1306.1 and 1306.2*
2 C *Table 1305.1*
3 A *1318.4*
4 C *1321.4*
5 B *1323.1.1*
6 A *1327.8.2 or Table 1305.1*
7 D *1306.2*
8 C *1324.5.5.2*
9 D *1324.5.1.1*
10 A *1312.2(1)*
11 A *1311.6*
12 C *1313.2(2)*
13 D *1323.13.2*
14 A *Table 1305.1*
15 C *1323.13.1*
16 D *1323.8*
17 A *1310.3*
18 B *1303.2*
19 A *1314.5(2)*
20 C *1303.4.1*
21 C *Table 1305.2*
22 B *1318.4*
23 C *1321.8.7*
24 B *1322.5(2)*
25 D *1321.8.9*
26 D *1321.9*
27 C *1321.7.2*
28 A *1323.3.1*
29 B *1314.10.2(1)*
30 B *1323.3.1*
31 A *1323.4.3*
32 D *1303.6*
33 D *1324.5.1.1*
34 C *1324.5.5.2*
35 C *1303.3*
36 A *1303.4.1*
37 B *1303.4.2*
38 B *1303.5*
39 A *1303.5*
40 C *1301.2(9)*
41 B *Table 1305.1*
42 B *1324.5.7.2*
43 C *1321.10.3*

CHAPTER 14 – FIRESTOP PROTECTION

1 D *1401.1*
2 D *1404.3*
3 B *1404.6 and 1405.6*
4 D *1404.4*
5 C *1406.3*
6 D *1405.5*
7 C *1406.2*
8 D *1406.1*
9 A *1404.6 and 1405.6*

CHAPTER 15 – ALTERNATE WATER SOURCES FOR NONPOTABLE APPLICATIONS

1 D *1501.2 Exception (1)*
2 D *1503.8.1(2)*
3 A *1503.9.5*
4 D *1505.4*
5 A *Table 1503.4 Note 9*
6 A *1505.9*
7 C *1504.5.7*
8 B *1504.2*
9 A *1504.5.2*
10 A *1503.4, and Table 1503.4*
11 B *Table 1503.4 Note 10*
12 B *1503.2.1*
13 B *1502.4*
14 A *1502.6*
15 A *1506.9 (leading to 601.3)*
16 A *1502.3.3(5)*
17 B *1503.8.1(1)*
18 C *1503.8.1(2)*
19 C *1503.9.1(9)*
20 D *1506.10.5*
21 D *1506.9 (leading to Table 601.3.2)*
22 C *1501.9.1*
23 B *Table 1504.2*
24 B *1503.8.1*
25 B *1504.4*
26 B *Table 1504.2*
27 B *1506.5*
28 D *1505.5*

CHAPTER 16 – NONPOTABLE RAINWATER CATCHMENT SYSTEMS

1	A	*1601.1*
2	A	*1602.1*
3	A	*1602.5*
4	A	*1603.2*
5	C	*1603.11*
6	A	*1603.3*
7	B	*1601.7 Exception (1)*
8	D	*1603.10*
9	A	*1603.13*
10	D	*1603.13*
11	C	*1603.18*
12	A	*1604.3*
13	B	*1604.2*
14	D	*1605.3.2(2)*
15	C	*1605.3.3(5)*
16	C	*1603.9*
17	A	*1603.8*
18	D	*1602.6*
19	B	*1603.12*
20	D	*1601.2 Exception (1)*
21	D	*Table 1603.4*
22	A	*1603.15*

CHAPTER 17 – REFERENCED STANDARDS

1	A
2	A
3	C
4	C
5	A
6	B
7	A
8	B
9	A
10	D
11	D
12	C
13	B
14	C
15	B
16	A
17	C
18	C
19	B
20	D
21	C
22	D
23	B
24	D
25	C
26	C

GENERAL EXAMINATION #1

1	A	*723.1*
2	C	*609.4*
3	C	*707.9*
4	C	*221.0*
5	B	*1208.15(5)*
6	A	*Table 313.3*
7	B	*Table 1215.2(7)*
8	D	*509.6.1(3)*
9	A	*1002.2*
10	A	*Tables 702.1 and 703.2*
11	B	*312.10*
12	B	*Table 313.3*
13	B	*Tables 702.1 and 703.2*
14	D	*707.9*
15	A	*801.3*
16	A	*509.7*
17	C	*1003.1 Ex.*
18	B	*Table 313.3*
19	A	*205.0*
20	B	*204.0*
21	D	*701.1(4)*
22	D	809.1
23	A	*701.2(3)*
24	D	*310.2*
25	A	*314.1*
26	A	*312.3 and Table 701.2*
27	A	*Table 313.3*
28	A	*609.2, 720.1, and Table 701.2*
29	B	*105.2.3*
30	B	*609.4*
31	B	*Table 702.1 Note 2 or Table 703.2 Note 2*
32	A	*Table 702.1*
33	A	*706.3*
34	C	*707.4*
35	C	*708.1*
36	C	*710.1*
37	C	*710.4*
38	C	*Table 703.2 Note 4*
39	D	*Table 703.2*
40	B	*Table 703.2*
41	B	*Tables 702.1 and 703.2*
42	B	*717.1 and Table 717.1*
43	A	*904.2 and Table 703.2 Note 6*
44	B	*906.2*
45	B	*907.1*
46	A	*807.3*
47	C	*803.3*
48	A	*809.1*
49	C	*811.1*
50	C	*908.1.1 and Table 702.1 Note 2*
51	C	*909.1*
52	C	*910.3*
53	A	*910.3*
54	A	*1001.2*
55	C	*1001.2*
56	B	*Table 1002.2*
57	C	*Table 1002.2 Note 2*
58	A	*1005.1*
59	C	*1014.2.1*
60	C	*Table 721.1 Note 3*
61	D	*Table 1210.3.5.1*
62	A	*609.5*
63	C	*402.6*
64	B	*906.5*
65	A	*712.2*
66	C	*402.10*
67	A	*701.4*
68	A	*407.5*
69	A	*402.5*
70	B	*804.1*
71	A	*408.6*
72	B	*509.6.2.2*
73	C	*408.8*
74	C	*603.5.1*
75	A	*603.5.2*
76	A	*608.2*
77	B	*608.3*
78	C	*609.1*
79	C	*608.5(6) and Appendix F - F501.8*
80	A	*707.4 Exception (1)*
81	C	*Table 703.2 Note 5*
82	A	*1003.1*
83	C	*1210.1.1*
84	C	*1210.9.2*
85	C	*1212.6*
86	D	*1208.3.1 and Table 1208.3.1*
87	A	*1213.4.1*
88	B	*1214.4*
89	B	*504.3.1*
90	D	*504.1(1)*
91	C	*507.13*
92	D	*509.6.1(1) and Figure 509.6.1*
93	A	*509.3.3.5*
94	C	*508.3.1*
95	D	*509.7*
96	D	*509.7.2(1)*

GENERAL EXAMINATION #2

1	D	104.4.3
2	B	104.2(2)
3	C	Table 703.2
4	B	904.1, Tables 702.1 and 703.2
5	D	807.1
6	A	702.2
7	D	1212.6
8	C	1214.4
9	B	509.8.1 and Table 509.8.1 Figure Clearance I
10	A	1002.1
11	D	609.4
12	D	610.6 leading to Appendix A- Table A104.4(1)
13	C	1213.3
14	C	Table 313.3
15	D	708.1
16	C	1215.4(1)
17	A	1208.2.1
18	D	1212.3.1(2)
19	B	701.2(6) and Table 701.2
20	C	707.4
21	B	Table 702.1 and Table 703.2
22	B	Table 702.1 and Table 703.2
23	B	1210.4.1(5)
24	D	706.1, 706.2, 706.3 and 706.4
25	D	712.2
26	A	1213.3
27	C	509.6.1(2)
28	B	910.7
29	B	1208.7.2
30	B	411.3
31	C	1214.4
32	A	608.2
33	C	608.2 and 608.3
34	D	710.1
35	A	603.2
36	C	706.3
37	D	Table 1014.2.1
38	C	608.1
39	D	1208.6.1.1
40	A	Table 703.2
41	A	510.1.11
42	B	506.4 and 506.5.3(3)
43	D	Table 1002.2
44	B	606.4
45	C	710.3(1)
46	B	Appendix A- A104.2
47	C	910.3
48	B	Table 313.3
49	D	803.3
50	A	803.3
51	A	810.1
52	D	Table 1002.2
53	B	Table 1002.2 Note (2)
54	B	1005.1
55	C	701.2(1) and 903.1(1)
56	C	507.5
57	B	609.3(2)
58	A	402.5
59	B	408.6
60	A	408.7(2)
61	B	713.2, Appendix H - H201.1, Table H201.1(1)
62	A	Appendix H - H701.6
63	C	604.5
64	D	Table 702.2(1)
65	B	Table 702.1
66	D	803.3
67	A	801.3.3
68	A	Table 702.1
69	C	Appendix H- H701.3
70	A	105.1
71	D	104.2(2)
72	B	409.2 and 420.4
73	A	904.2 and Table 703.2 Note 6
74	C	905.3
75	A	906.1
76	B	610.7(3) and 610.8(2)
77	A	Table 1208.3.1, Figure 1215.1.1, Table 1215.2(1)
78	A	402.6.1
79	B	1213.4.1
80	C	705.2.1
81	C	Table 707.1
82	C	910.4 and Appendix B - B101.6.1
83	B	505.1
84	D	504.1(2)
85	A	508.4
86	D	1003.1
87	A	804.1
88	B	706.3
89	C	718.1 Exception
90	B	722.5
91	C	604.13
92	A	608.5
93	B	606.2
94	D	710.4
95	B	1005.1
96	B	811.1
97	C	506.2.1
98	D	1603.8

GENERAL EXAMINATION #3

1	D	104.2
2	B	910.2
3	D	Table 703.2, Note 4
4	C	312.10
5	A	310.5
6	B	314.1
7	C	Table 313.3
8	A	712.2
9	D	Table 702.1
10	D	701.2(2)
11	A	701.2(4)
12	D	706.4
13	A	707.4
14	B	Table 707.1
15	B	708.1
16	B	710.3(2)
17	C	Table 702.1 and 703.2
18	B	Table 703.2
19	D	906.7
20	A	905.1
21	A	803.3
22	A	810.1
23	C	910.4 and Appendix B - B101.5 and B101.6.1
24	B	1001.2
25	B	1001.2
26	C	Table 1002.2
27	A	1005.1
28	B	402.6.1
29	A	420.4
30	D	Appendix A- A104.1
31	B	301.2.2
32	A	705.2, 705.2.1 and 705.2.2
33	D	315.1
34	C	402.10
35	B	402.5
36	D	408.7
37	B	603.5.1
38	A	606.2
39	A	608.2
40	B	610.8(6) and Table 610.4 Note 2
41	D	717.1
42	B	719.1
43	B	722.1
44	A	1210.1.1
45	B	1214.4
46	C	1212.3.1(2)
47	C	506.3(1)
48	A	507.13
49	B	150 divided by 8 = 18.75
50	B	705.9.2
51	B	1203.3.1
52	B	610.8(2)
53	D	Table 703.2 Note 4
54	D	710.4
55	A	909.1
56	B	1001.2
57	A	901.2
58	C	705.2.1
59	D	408.9
60	D	603.5.12
61	A	610.10 and Table 610.10
62	A	Appendix A - A107.1
63	B	713.2, Appendix H - H201.1, Table H201.1(1)
64	B	509.10.5
65	D	1210.3.4
66	B	408.6
67	B	719.1
68	D	710.6
69	A	310.1
70	A	Table 313.3
71	C	105.2.3
72	B	723.1
73	D	701.2, Table 701.2
74	A	Table 702.1
75	C	Table 702.1, Note 5
76	A	Appendix B - B101.2
77	B	509.6.1(1)(a)
78	B	509.7
79	C	1214.3
80	C	1213.4.1
81	A	1203.3.1
82	C	Appendix H - H801.3
83	D	Appendix D - Table D101.1
84	C	Table 603.2
85	B	402.4, 402.6
86	D	1003.1
87	A	508.3.1
88	D	1003.1
89	A	Table 1002.2 Note 2
90	B	1004.1
91	A	803.3
92	A	804.1
93	C	907.1
94	B	801.3.1
95.	D	204.0
96.	A	814.5 and 814.6
97	D	904.1 Exception

PLUMBING MATHEMATICS EXAMINATION #1
ANSWERS AND EXPLANATIONS

1. (B) Atmospheric pressure is a unit of pressure equal to 14.7 pounds per square inch (101.3 KPa). The pressure exerted by the air on the earth's surface at sea level is about one atmosphere.

2. (C) One cubic foot contains:
A x B x C = 12" x 12" x 12" = 1,728 cu. in.

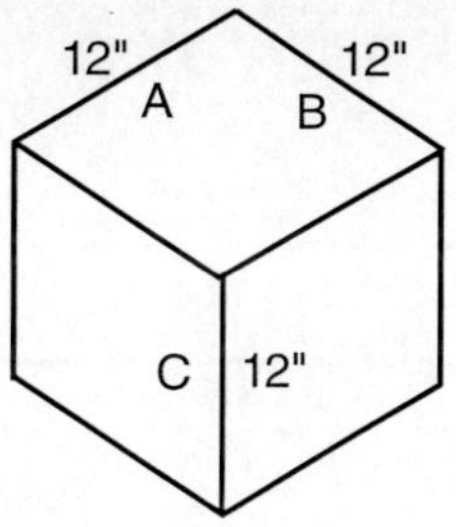

3. (A) One cubic foot contains = 1,728 cu. in.
One gallon of water contains = 231 cu. in.
One cubic foot of water = $\frac{1,728}{231}$ = 7.4805 gals.

4. (C) 1 gal. weighs 8.3453 lbs.
1 gal. (8.3453) x 7.4805 gal. = 62.427 lbs./cu. ft.
$\frac{62.427 \text{ lbs./cu. ft.}}{144 \text{ in./sq. ft.}}$ = 0.4335 psi $\frac{0.4335}{12}$ = 0.0361 psi

5. (A) 1 gal. water weighs 8.34 lbs.

6. (C) Pressure at base of column water 1 ft. high = 0.4335 psi
Pressure at base of column 200 ft. high = 200 x .4335 = 86.7 psi
Therefore requires a pressure of 86.7 psi to raise column of water 200 ft. high

7. (A) $\frac{25 \text{ in.}}{100 \text{ ft.}}$ = 1/4 in./ft.

8. (C) 1 gal. 8.3453 lbs. x 7.4805 gal./cu. ft. = 62.427 lbs./cu. ft.

9. (B) (˚F -32)/1.8 = ˚C
(212˚F -32)/1.8 = ˚C

180/1.8 = 100 ˚C

10. (A) 12" x 12" = 144 sq. in.

11. (C) Any pipe diameter will do. For the example use a 2" dia.
Area of 2" dia. pipe = 3.1416 sq. in.
Double the diameter = 2" x 2" = 4" dia.
Area of 4" dia. pipe = 12.567 sq. in.
$\frac{\text{Area of 4"}}{\text{Area of 2"}} = \frac{12.567 \text{ sq. in.}}{3.1416}$ = 4 times

12. (D) Scale 1/8" = 1 ft.
$8 \times 5 = 40$ ft.

13. (A) Area of Pipe: $\pi d^2/4$ or πr^2
$A = 2^2\pi/4$
$A = 3.14\ in^2$
Or
$A = \pi 1^2$
$A = 3.14\ in^2$

14. (C) 4' = 48"
Constant = $\sqrt{2}$
48" x $\sqrt{2}$ = 67.88
67.88/12 = 5.66 ft (approximately 5'8")

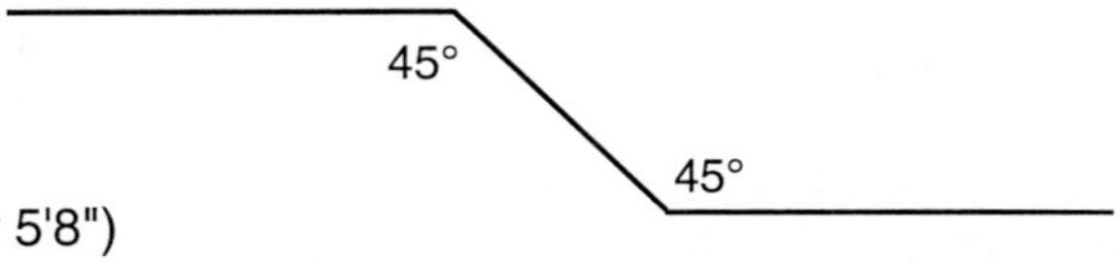

15. (C)
1. Pressure of bottom of column of water 100 ft. high = 100 ft. x .434 psi = 43.4 psi
2. Area of 4 in. pipe = 12.56 sq. in.
3. Pressure over total area = (12.56 in^2) x (43.4 lbs./in^2) = 545 lbs.

16. (D) Formula for area of a circle = $\pi d^2/4$
d^2 = Diameter of circle. Exponent 2 means squared.

17. (C) Formula for circumference = πd
Formula for area = πr^2 or $\pi d^2/4$

Using the above formulas on a 4 in. circle
$C = \pi d = (3.14159)(4) = 12.567$ in.

$A = \pi r^2 = (3.14159)(4) = 12.567$ sq. in.

18. (C) Area of square = 196 ft.^{2}
Formula for root (side) of a square =
Side = $\sqrt{196} = 14$

19. (D)

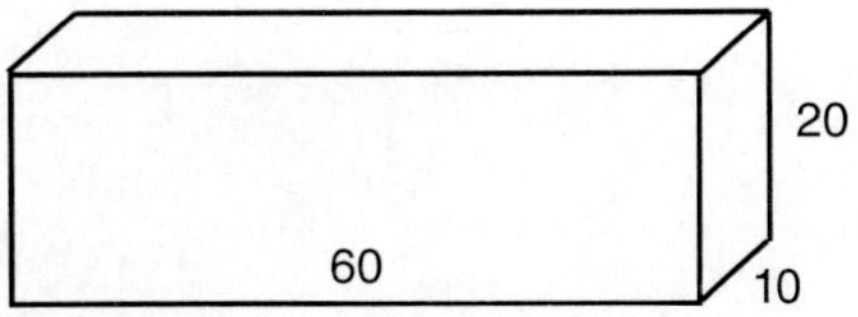

2 sides 60 x 20 x 2 = 2,400 sq. ft.
2 sides 60 x 10 x 2 = 1,200 sq. ft.
2 sides 20 x 10 x 2 = 400 sq. ft.
Total 4,000 sq. ft.

20. (A) Formula for volume of a cylinder $V = (\pi r^2)h$
Formula: $V = (\pi r^2)h$
$V = (2^2\pi) \times 20$
$V = 251.33\ ft^3$
Volume in gallons: $(2^2\pi) \times 20 \times 7.4805 \approx 1880$ gallons

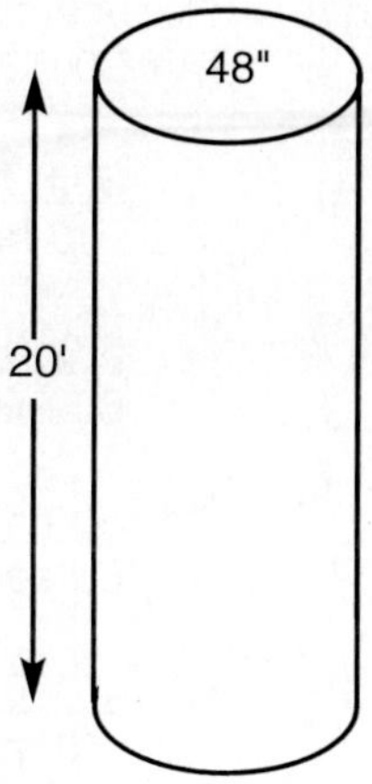

21. (B) Area of 2" dia. pipe = 3.14 sq. in.
(4) - 2" dia. pipe x 3.14 sq. in. = 12.56 sq. in.
Area of 4" dia. pipe = 12.56 sq. in.
Area of (4) - 2" dia. pipe = Area of 4" dia. pipe

22. (D)

1.

8 ft.	6 in.
6 ft.	7 in.
10 ft.	4 in.
9 ft.	2 in.
4 ft.	4 in.
37 ft.	23 in.

2. 37 ft. 23 in. = 38 ft. 11 in.

23. (A) Formula: $a^2 + b^2 = c^2$
$A = 3\ \ A^2 = 9$
$B = 4\ \ B^2 = 16$
$9 + 16 = 25$
$\sqrt{25} = 5$

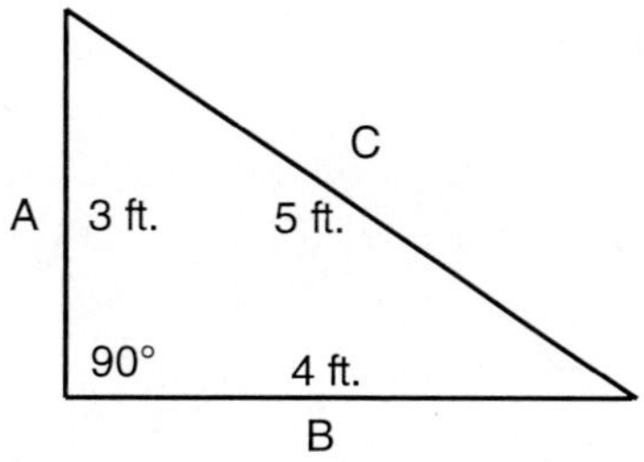

24. (D) Formula: $l = 1/2(2\pi r) + 2r$

$r = 2.5$ft.
$l = 1/2(2\pi \times 2.5) + (2 \times 2.5)$
$l = 2.5\pi + 5$
$l = 12.85$ ft

For 4 straps, $4(2.5\pi + 5) = 51.42$ ft.

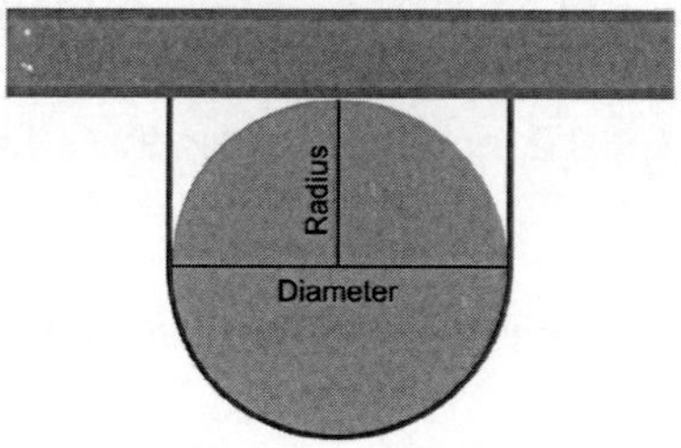

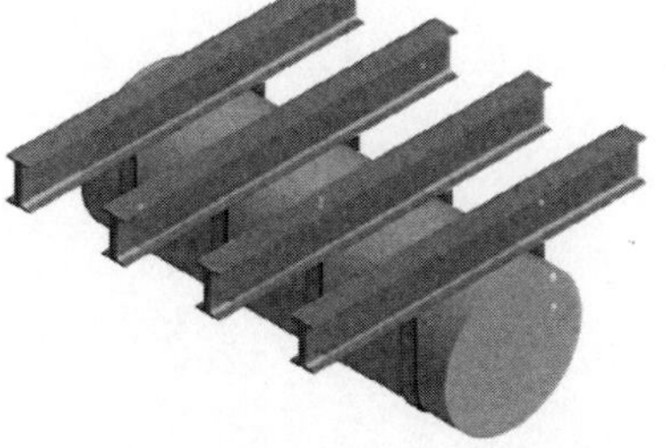

25. (D) Reduce answers to flow per second

(A) 30,000 gal. in 10 hours =

$\frac{30{,}000 \text{ gal.}}{10 \text{ hrs.}}$ = 3,000 gal./60 min./hr.

$\frac{3{,}000 \text{ gal.}}{60 \text{ min./hr.}}$ = 50 gal./min.

$\frac{50 \text{ gal./min.}}{60 \text{ sec./min.}}$ = 0.83 gal./sec.

(B) $\frac{1{,}000 \text{ gal./hr.}}{60 \text{ min./hr.}}$ = 16.66 gal./min.

$\frac{16.66 \text{ gal./min.}}{60 \text{ sec./ min.}}$ = 0.277 gal./sec.

(C) $\frac{20 \text{ gal./min.}}{60 \text{ sec./min.}}$ = 0.33 gal./sec.

(D) 1 gal./sec.

Answer (D) highest rate of floor

26. (B) 864 x 7.48 = 6,462.72 gallons

27. (A) 9.2 inches

πr^2 = cross-sectional area of tank

$20^2\pi = 1256.637061$

1256.637061 in^2 x 1 inch = 1256.637061

Therefore 1 inch of water in the tank will contain 1,256.637061 cubic inches.

There are 231 cubic inches in a gallon.

1,256.637061 ÷ 231 = 5.439987279 gallons in one inch of water in the tank

50 (minimum gallon discharge) ÷ 5.439987279 (gallons in 1 inch of water in the tank)

= 9.2 inches between the start and stop float

28. (C) 8 inches x .036125 psi/inch (Useful Tables in UPC Study Guide) .289 psi

29. (B) Double the diameter, quadruple the area.

(Double the radius and then square it increases area by a factor of 4.)

30. (B) Formula: $A = \pi d^2/4$

$A = \pi 4^2/4$

$A = 4\pi$

$A = 12.57 \text{ in}^2$

PLUMBING MATHEMATICS EXAMINATION #2
ANSWERS AND EXPLANATIONS

1. Formula: $A = 2(hw + hl + wl)$
   If l=25, w=15, h=8
   Then $A = 2[(8x15) + (8x25) + (15x25)]$
   A = <u>1390 ft^2</u>

2. Formula: V = L x W x H
   V = 8x15x25
   V = <u>3000 ft^3</u>

3. Formula: $A = 2(hw + hl + wl)$. Change units to feet.
   If l=9.25 ft., w=4.5 ft., h=22.75 ft., then
   $A = 2[(22.75x4.5) + (22.75x9.25) + (4.5x9.25)]$
   $A = 2(102.375 + 210.4375 + 41.625)$
   A = <u>708.88 ft^2</u> or <u>102,078.72 in^2</u>

4. Formula: V= L x W x H
   V = 9.25x4.5x22.75
   V= <u>946.97ft^3</u> or <u>1,636,362 in^3</u>

5. Formula: $V = (\pi r^2)h$
   $V= (1.75^2\pi) \times 15.25$
   V = <u>146.72 ft^3</u> or <u>253,535.95 in^3</u>

6. Do not use the rounded answer in 5. Look in the Useful Tables for properties of water. If the volume of a cylinder is $(\pi r^2)h$, then add to the formula the weight of water per cubic foot. Cubic foot of water = 62.427 lbs.
   Then $(1.75^2\ \pi) x15.25 \times 62.427$ = <u>9,159.43 lbs</u>

7. Formula: A = 1/2(bh)
   A = (8 x 15)/2
   A = <u>60 in^2</u>

8. Formula: A = 1/2(bh)
   A = (15 x 15)/2
   A = <u>112.5 in^2</u>

9. Formula: $a^2 + b^2 = c^2$ where a and b are sides.
   $15^2 + 15^2 = C^2$
   $450 = C^2$
   $C = \sqrt{450}$
   C = <u>21.21 in</u>
   Since this is an isosceles right triangle use Formula $H = s\sqrt{2}$
   $H = 15\sqrt{2}$
   H = <u>21.21 in.</u>

10. Formula: $a^2 + b^2 = c^2$
$5^2 + 10^2 = C^2$
$125 = c^2$
$c = \sqrt{125}$
c = <u>11.18 in.</u>

11. Formula $A = \pi r^2$
$A = \pi 10^2$
$A = 100\pi$
A = <u>314.16 ft²</u>

12. Formula $A = s^2$
$324 = s^2$
$\sqrt{324} = s$
s = <u>18 in.</u>

13. Formula: $A = 2\pi r^2 + 2\pi rh$.
Change units to feet.
$A = 2\pi 2^2 + 2\pi(2x20)$
$A = 8\pi + 80\pi$
A = <u>276.46 ft²</u>

14. Formula: A = 1/2(bh)
A = (20 x 10)/2
A = <u>100 in²</u>

15. Formula: $A = \pi r^2$
$A = 7.5^2\pi$
$A = 176.72\ ft^2$

16. Formula: $a^2 + b^2 = c^2$
$12^2 + 12^2 = c^2$
$144+144= c^2$
$288 = c^2$
$c = \sqrt{288}$
C = <u>16.97 in.</u>
Since this is an isosceles right triangle use Formula $H = s\sqrt{2}$.
$H = 12\sqrt{2}$
H = <u>16.97 in.</u>

17. Formula: A = 1/2(bh)
A = (12 x 12)/2
A = <u>72 in²</u>

18. Formula: $a^2 + b^2 = c^2$
$8^2+16^2 = c^2$
$64 + 256 = c^2$
$c^2 = 320$
$c = \sqrt{320}$
c = <u>17.89 in.</u>

19. Formula: $A = s^2$
$361 = s^2$
$s = \sqrt{361}$
s = <u>19 in.</u>

20. Formula: $A = 2\pi r^2 + 2\pi rh$.
Change units to feet.
$A = 2\pi 1.5^2 + 2\pi(1.5x11)$
$A = 4.5\pi + 33\pi$
A = <u>117.81 ft^2</u>

21. Formula: A = 2(hw +hl + wl)
If l=7, w=18, h=30
Then 2[(30x18) + (30x7) + (18x7)] = <u>1,752 ft^2</u>

22. The equation can be constructed from two parts of a circle, the circumference and radius. The first part is 1/2 the circumference. The second part is twice the radius.
Equation: $l = 1/2(2\pi r) + 2r$
r = 1.5 ft.
$l = 1/2(2\pi \times 1.5) + (2 \times 1.5)$
$l = 1.5\pi + 3$
l = 7.71 ft.
For 3 straps, $3(1.5\pi + 3)$ = <u>23.14 ft.</u>

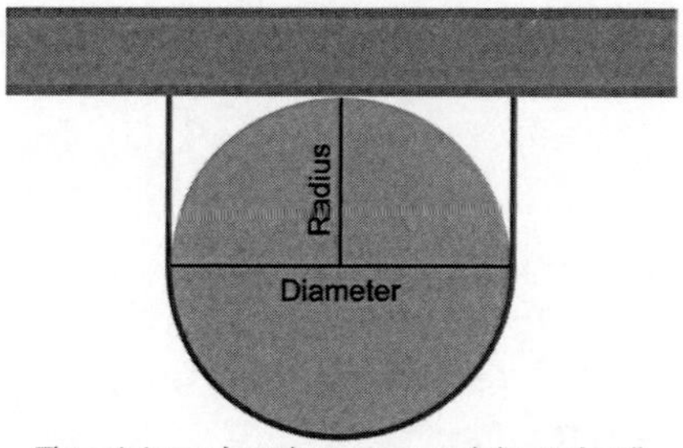

The minimum length strap around the tank will encompass the lower ½ of the circumference and leaving contact of the tank at the mid point will travel vertically to one of the steel beams. Thus, the length of each strap is ½ the circumference plus 2 radii (or the diameter). Formula for computation, ½ C + D

23. The equation can be constructed from two parts of a circle, the circumference and radius. The first part is 1/2 the circumference. The second part is twice the radius.
Equation: $l =1/2(2\pi r) + 2r$
r = 3.75 ft.
$l =1/2(2\pi \times 3.75) + (2 \times 3.75)$
$l = 3.75\pi + 7.5$
l = 19.28 ft.
For 5 straps, $5(3.75\pi + 7.5)$ = <u>96.41 ft.</u>

24. For a 45-degree offset, the offset is the hypotenuse of an isosceles right triangle.
Formula: $L = s\sqrt{2}$
$L = 87\sqrt{2}$
L = <u>123.04 in.</u>

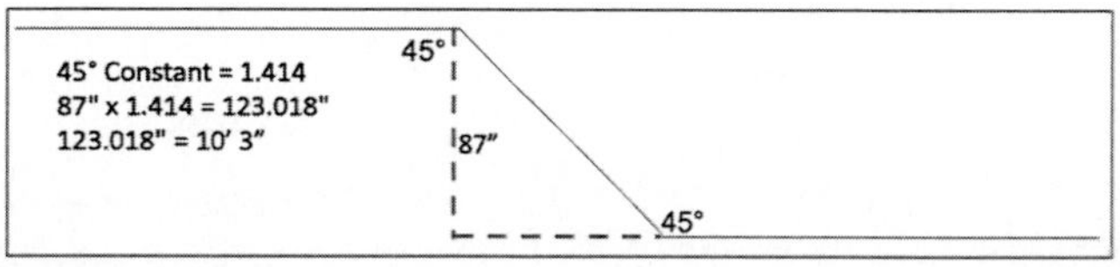

TRADE EXAMINATIONS

Drawing 1

Pipe Section	Fixture Units	Pipe Size	Pipe Section	Fixture Units	Pipe Size	Pipe Section	Fixture Units	Pipe Size
1	1	1-1/4"	**10**	3	2"	**19**	2	1-1/2"
2	1	2"	**11**	9	3"	**20**	3	2"
3	2	2"	**12**	3	3"	**21**	3	1-1/2"
4	3	2"	**13**	3	3"	**22**	6	2"
5	3	3"	**14**	12	3"	**23**	8	2"
6	3	3"	**15**	2	1-1/2"	**24**	3	1-1/2"
7	6	3"	**16**	2	2"	**25**	3	2"
8	2	1-1/2"	**17**	14	3"	**26**	11	3"
9	1	2"	**18**	14	3"	**27**	14	3"

Drawing 2

Pipe Section	Fixture Units	Pipe Size	Pipe Section	Fixture Units	Pipe Size	Pipe Section	Fixture Units	Pipe Size	Pipe Section	Fixture Units	Pipe Size	Pipe Section	Fixture Units	Pipe Size
1	2	1-1/2"	**8**	10	3"	**15**	1	1-1/4"	**22**	22	3"	**29**	5	2"
2	2	2"	**9**	2	2"	**16**	5	2"	**23**	22	3"	**30**	10	2"
3	1	1-1/4"	**10**	12	3"	**17**	2	1-1/2"	**24**	2	1-1/2"	**31**	2	1-1/2"
4	2	1-1/2"	**11**	2	2"	**18**	2	3"	**25**	2	1-1/2"	**32**	6	2"
5	4	2"	**12**	2	2"	**19**	5	3"	**26**	4	1-1/2"	**33**	16	2"
6	3	3"	**13**	2	1-1/2"	**20**	10	3"	**27**	1	1-1/4"	**34**	2	1-1/2"
7	6	3"	**14**	4	2"	**21**	10	3"	**28**	5	1-1/2"	**35**	2	1-1/2"

Drawing 3

Pipe Section	Fixture Units	Pipe Size	Pipe Section	Fixture Units	Pipe Size	Pipe Section	Fixture Units	Pipe Size	Pipe Section	Fixture Units	Pipe Size	Pipe Section	Fixture Units	Pipe Size
1	4	3"	**9**	1	1-1/4"	**17**	8	3"	**25**	18	2"	**33**	18	2"
2	8	3"	**10**	13	4"	**18**	16	3"	**26**	2	1-1/2"	**34**	8	3"
3	2	2"	**11**	1	1-1/4"	**19**	2	1-1/2"	**27**	18	2"	**35**	26	3"
4	10	3"	**12**	14	4"	**20**	18	3"	**28**	8	2"	**36**	2	1-1/2"
5	0	2"	**13**	8	3"	**21**	20 gpm	3"	**29**	18	2"	**37**	0	1-1/4"
6	10	4"	**14**	8	3"	**22**	20 gpm	3"	**30**	0	1-1/4"	**38**	2	1-1/2"
7	2	2"	**15**	0	2"	**23**	20 gpm	3"	**31**	18	2"	**39**	2	1-1/2"
8	12	4"	**16**	8	3"	**24**	54	4"	**32**	8	2"	**40**	1	1-1/4"
												41	1	1-1/4"

Drawing 4

Pipe Section	Fixture Units	Pipe Size	Pipe Section	Fixture Units	Pipe Size	Pipe Section	Fixture Units	Pipe Size	Pipe Section	Fixture Units	Pipe Size	Pipe Section	Fixture Units	Pipe Size
1	4	3"	**13**	0	2"	**25**	0	2"	**36**	20	2"	**47**	2	1-1/2"
2	8	3"	**14**	6	2"	**26**	12	3"	**37**	0	1-1/4"	**48**	4	1-1/2"
3	0	2"	**15**	2	2"	**27**	8	2"	**38**	20	2'	**49**	4	1-1/2"
4	8	3"	**16**	8	2"	**28**	20	3"	**39**	8	2'	**50**	4	2"
5	4	2"	**17**	1	1-1/4"	**29**	20 gpm	3"	**40**	20	2"	**51**	6	2"
6	12	4"	**18**	2	1-1/2"	**30**	20 gpm	3"	**41**	4	2"	**52**	4	2"
7	4	2"	**19**	2	3"	**31**	20 gpm	3"	**42**	20	2"	**53**	8	2"
8	16	4"	**20**	10	3"	**32**	50	4"	**43**	2	1-1/2"	**54**	0	1-1/4"
9	2	2"	**21**	4	3"	**33**	66	4"	**44**	2	1-1/2"	**55**	8	3"
10	4	2"	**22**	4	3"	**34**	20	2"	**45**	0	1-1/4"	**56**	24	2"
11	2	2"	**23**	8	3"	**35**	8	2"	**46**	2	1-1/2"	**57**	32	3"
12	6	2"	**24**	12	3"									

Drawing 5

Pipe Section	Fixture Units	Pipe Size	Pipe Section	Fixture Units	Pipe Size	Pipe Section	Fixture Units	Pipe Size	Pipe Section	Fixture Units	Pipe Size	Pipe Section	Fixture Units	Pipe Size
1	1	1-1/4"	**13**	11	3"	**25**	0	4"	**37**	59	4"	**48**	4	2"
2	1	2"	**14**	2	2"	**26**	4	2"	**38**	0	4"	**49**	6	3"
3	3	3"	**15**	2	2"	**27**	2	2"	**39**	59	4"	**50**	2	1-1/2"
4	4	3"	**16**	2	2"	**28**	4	2"	**40**	11	2"	**51**	8	3"
5	2	1-1/2"	**17**	1	1-1/4"	**29**	6	3"	**41**	1	1-1/4"	**52**	19	3"
6	2	2"	**18**	1	2"	**30**	3	3"	**42**	11	2"	**53**	4	2"
7	6	3"	**19**	3	3"	**31**	6	3"	**43**	6	2"	**54**	23	4"
8	3	2"	**20**	4	3"	**32**	10	3"	**44**	11	2"	**55**	2	1-1/2"
9	3	2"	**21**	6	3"	**33**	1	1-1/4"	**45**	4	1-1/2"	**56**	3	1-1/2"
10	9	3"	**22**	2	1-1/2"	**34**	11	3"	**46**	11	2"	**57**	5	1-1/2"
11	2	1-1/2"	**23**	2	1-1/2"	**35**	20 gpm	2"	**47**	2	1-1/2"	**58**	2	1-1/2"
12	2	2"	**24**	8	3"	**36**	48	4"						

Drawing 6

Demand from Table 610.3 in WSFU					
	Building Supply			Branches	
Fixture	WSFU	QTY	Total WSFU	Cold FU @100%	Hot FU @100%
WC, 1.6 GPF (Gravity Tank)	2.5	2	5	5	-
Lavatory	1	2	2	2	2
Bathtub	4	1	4	4	4
Shower	2	1	2	2	2
Kitchen Sink with DW	1.5	1	1.5	1.5	1.5
Laundry Sink	-	-	-	-	-
Clothes Washer	-	-	-	-	-
Bar Sink	-	-	-	-	-
Hose Bibb	2.5	1	2.5	2.5	-
Additional Hose Bibbs	-	-	-	-	-
Wash Fountain, Circular Sink	-	-	-	-	-
Total Demand at Meter			17	-	-
Total for Hot and Cold Demands				17	9.5

Section	FU	Size	Section	FU	Size
Meter	17	3/4"	16	4	3/4"
2	17	1"	17	2	1/2"
3	2	1/2"	18	6	3/4"
4	4	3/4"	19	1	1/2"
5	6	3/4"	20	1	1/2"
6	1	1/2"	21	2	1/2"
7	7	3/4"	22	8	3/4"
8	1	1/2"	23	1.5	1/2"
9	8	3/4"	24	9.5	3/4"
10	2.5	1/2"	25	9.5	3/4"
11	10.5	3/4"	26	17*	1"
12	2.5	1/2"	27	2.5	1/2"
13	13	1"			
14	1.5	1/2"			
15	14.5	1"			

Net Water Press = 35
Length Column = 100
Meter Size = 3/4"
1/2" = 3 fu
3/4" = 12 fu
1" = 21 fu

Drawing 7

Demand from Table 610.3 in WSFU					
	Building Supply			Branches	
Fixture	WSFU	QTY	Total WSFU	Cold FU @100%	Hot FU @100%
WC, 1.6 GPF (Gravity Tank)	2.5	3	7.5	7.5	-
Lavatory	1	3	3	3	3
Bathtub	4	2	8	8	8
Shower	2	1	2	2	2
Kitchen Sink with DW	1.5	1	1.5	1.5	1.5
Laundry Sink	1.5	1	1.5	1.5	1.5
Clothes Washer	-	-	-	-	-
Bar Sink	-	-	-	-	-
Hose Bibb	2.5	1	2.5	2.5	-
Additional Hose Bibbs	1	1	1	1	-
Wash Fountain, Circular Sink	-	-	-	-	-
Total Demand at Meter			27	-	-
Total for Hot and Cold Demands				27	16

Section	FU	Size	Section	FU	Size	Section	FU	Size
Meter	27	3/4"	16	8.5	3/4"	31	1	1/2"
2	27	1-1/4"	17	16	1"	32	27*	1-1/4"
3	1.5	1/2"	18	16	1"	33	2.5	1/2"
4	1	1/2"	19	2.5	1/2"	34	2.5	1/2"
5	2.5	1/2"	20	18.5	1"	35	5	3/4"
6	1	1/2"	21	2	1/2"	36	2.5	1/2"
7	3.5	3/4"	22	20.5	1"	37	7.5	3/4"
8	4	3/4"	23	4	3/4"	38	27*	1-1/4"
9	7.5	3/4"	24	1	1/2"	39	1.5	1/2"
10	1.5	1/2"	25	5	3/4"	40	2.5	1/2"
11	1	1/2"	26	1	1/2"	41	4	3/4"
12	2.5	1/2"	27	6	3/4"	42	1.5	1/2"
13	4	3/4"	28	4	3/4"	43	5.5	3/4"
14	6.5	3/4"	29	10	3/4"			
15	2	1/2"	30	27*	1-1/4"			

Net Water Press = 43
Length Column = 100
Meter Size = 3/4"
1/2" = 3 fu
3/4" = 13 fu
1" = 21 fu
1-1/4" = 28 fu

Drawing 8

Demand from Table 610.3 in WSFU					
	Building Supply			Branches	
Fixture	WSFU	QTY	Total WSFU	Cold FU @100%	Hot FU @100%
WC, 1.6 GPF (Flushometer	2.5	6	15	15	-
Urinal 1.0 GPF (Flushometer)*	20	1	20	20	-
Urinal 1.0 GPF (Flushometer)*	20	1	15	15	-
* See Footnote 7	-	-	-	-	-
Lavatory	1	4	4	4	4
Hose Bibb	2.5	1	2.5	2.5	-
Additional Hose Bibbs	1	1	1	1	-
Wash Fountain, Circular Sink	-	-	-	-	-
Total Demand at Meter			57.5	-	-
Total for Hot and Cold Demands				57.5	4

Section	FU	Size	Section	FU	Size
Meter	57.5	1"	21	1	1/2"
2	57.5	1-1/4"	22	2	1/2"
3	2.5	1/2"	23	54.5	1-1/4"
4	2.5	1/2"	24	1	1/2"
5	5	1/2"	25	1	1/2"
6	2.5	1/2"	26	2	1/2"
7	7.5	3/4"	27	56.5	1-1/4"
8	2.5	1/2"	28	1	1/2"
9	2.5	1/2"	29	1	1/2"
10	5	1/2"	30	2	1/2"
11	12.5	3/4"	31	1	1/2"
12	20	1"	32	1	1/2"
13	2.5	1/2"	33	2	1/2"
14	22.5	1"	34	4	1/2"
15	35	1"	35	4	1/2"
16	20	1"	36	2.5	1/2"
17	2.5	1/2"	37	6.5	3/4"
18	22.5	1"			
19	52.5	1-1/4"			
20	1	1/2"			

Net Water Press = 50
Length Column = 100
Meter Size = 1"
1/2" = 5 fu
3/4" = 17 fu
1" = 36 fu
1-1/4" = 67 fu

Drawing 9

Demand from Table 610.3 in WSFU					
	Building Supply			Branches	
Fixture	WSFU	QTY	Total WSFU	Cold FU @100%	Hot FU @100%
WC, 1.6 GPF (Flushometer)*	40	1	40	40	-
WC, 1.6 GPF (Flushometer)*	40	1	30	30	-
WC, 1.6 GPF (Flushometer)*	40	1	20	20	-
WC, 1.6 GPF (Flushometer)*	40	1	15	15	-
WC, 1.6 GPF (Flushometer)*	40	3	30	30	-
Urinal 1.0 GPF (Flushometer)*	20	1	20	20	-
Urinal 1.0 GPF (Flushometer)*	20	1	15	15	-
* See Footnote 7	-	-	-	-	-
Lavatory	1	2	2	2	2
Shower	2	4	8	8	8
Hose Bibb	2.5	1	2.5	2.5	-
Additional Hose Bibbs	1	2	2	2	-
Wash Fountain, Circular Spray	4	2	8	8	8
Total Demand at Meter			193	-	-
Total for Hot and Cold Demands				192.5	18

Section	FU	Size	Section	FU	Size	Section	FU	Size
Meter	192.5	2"	21	40	1-1/2"	41	41.5	1-1/2"
2	192.5	2"	22	88	2"	42	1	1/2"
3	2	1/2"	23	2	1/2"	43	42.5	1-1/2"
4	2	1/2"	24	90	2"	44	1	1/2"
5	4	3/4"	25	2	1/2"	45	43.5	1-1/2"
6	4	3/4"	26	92	2"	46	40	1-1/2"
7	8	3/4"	27	4	3/4"	47	40	1-1/2"
8	4	3/4"	28	96	2"	48	70	1-1/2"
9	12	1"	29	4	3/4"	49	40	1-1/2"
10	1	1/2"	30	100	2"	50	90	2"
11	13	1"	31	2.5	3/4"	51	40	1-1/2"
12	1	1/2"	32	102.5	2"	52	105	2"
13	14	1"	33	2.5	3/4"	53	40	1-1/2"
14	2	1/2"	34	2	1/2"	54	115	2"
15	16	1"	35	4.5	3/4"	55	2.5	3/4"
16	2	1/2"	36	2	1/2"	56	117.5	2"
17	18	1"	37	6.5	3/4"	57	159.5	2"
18	18	1"	38	20	1"			
19	40	1-1/2"	39	26.5	1-1/4"			
20	58	1-1/2"	40	20	1"			

Net Water Press = 40
Length Column = 150
Meter Size = 2"
1/2" = 2 fu
3/4" = 9 fu
1" = 20 fu
1-1/4" = 38 fu
1-1/2" = 80 fu
2" = 217 fu

Drawing 10

Demand from Table 610.3 in WSFU					
	Building Supply			Branches	
Fixture	WSFU	QTY	Total WSFU	Cold FU @100%	Hot FU @100%
WC, 1.6 GPF (Gravity Tank)	2.5	4	10	10	-
Lavatory	1	3	3	3	3
Bathtub	4	1	4	4	4
Shower	2	3	6	6	6
Kitchen Sink with DW	1.5	1	1.5	1.5	1.5
Laundry Sink	1.5	1	1.5	1.5	1.5
Clothes Washer	4	1	4	4	4
Bar Sink	-	-	-	-	-
Hose Bibb	2.5	1	2.5	2.5	-
Additional Hose Bibbs	1	1	1	1	-
Wash Fountain, Circular Sink	-	-	-	-	-
Total Demand at Meter			33.5	-	-
Total for Hot and Cold Demands				33.5	20

Section	FU	Size	Section	FU	Size	Section	FU	Size
Meter	33.5	1"	21	1	1/2"	41	3.5	1/2"
2	33.5	1"	22	8	3/4"	42	33.5*	1"
3	1	1/2"	23	20	1"	43	2	1/2"
4	4	1/2"	24	20	1"	44	2	1/2"
5	1	1/2"	25	2.5	1/2"	45	4	1/2"
6	5	1/2"	26	22.5	1"	46	2	1/2"
7	6	3/4"	27	1	1/2"	47	6	3/4"
8	2	1/2"	28	23.5	1"	48	2.5	1/2"
9	8	3/4"	29	1.5	1/2"	49	8.5	3/4"
10	2	1/2"	30	25	1"	50	2.5	1/2"
11	2	1/2"	31	4	1/2"	51	11	3/4"
12	4	1/2"	32	29	1"	52	1	1/2"
13	12	3/4"	33	1.5	1/2"	53	12	3/4"
14	1.5	1/2"	34	30.5	1"	54	33.5*	1"
15	1.5	1/2"	35	2.5	1/2"	55	2.5	1/2"
16	1.5	1/2"	36	33	1"			
17	4	1/2"	37	4	1/2"			
18	5.5	3/4"	38	33.5*	1"			
19	1.5	1/2"	39	2.5	1/2"			
20	7	3/4"	40	1	1/2"			

Net Water Press = 60
Length Column = 100
Meter Size = 1"
1/2" = 5 fu
3/4" = 17 fu
1" = 36 fu

Drawing 11

From Table 610.4
Pressure Range - 46 to 60 psi 200 foot column

Pipe Size	Max. Fixture Units
1"	25
3/4"	11
1/2"	3

Demand from Table 610.3 in WSFU

	Building Supply			Branches - 75% Method		
Fixture	**WSFU**	**QTY**	**Total WSFU**	**Cold WSFU (each outlet)**	**Hot WSFU (each outlet)**	**Total Hot WSFU**
WC, 1.6 gpf	2.5	2	5.0	2.5		
Lavatory	1.0	2	2.0	0.75	0.75	1.5
Shower	2.0	1	2.0	1.5	1.5	1.5
Combination Bath/Shower	4.0	1	4.0	3.0	3.0	3.0
Laundry Sink	1.5	1	1.5	1.125	1.125	1.125
Clothes Washer	4.0	1	4.0	3.0	3.0	3.0
Kitchen sink w/Dishwasher	1.5	1	1.5	1.125	1.125	1.125
1st. Hose Bibb	2.5	1	2.5	2.5		
Additional Hose Bibbs	1.0	1	1.0	1.0		
Total Demand at water meter			23.5			
Total Demand at water heater						11.25

Hose bibb sizing. Each pipe section serving only one hose bibb is assigned a fixture unit value of 2.5. Section B serves on hose bibb with a rating of 2.5 plus one "additional" hose bibb with a fixture unit value 1.0. The total hose bibb fixture unit value at section B is 3.5. See Note 8 at bottom of the Table 610.3.

Pipe Section	Fixture Units	Pipe Size
A meter	23.5	1"
B	23.5	1"
C	5.5	¾"
D	8.0	¾"
E	8.75	¾"
F	9.5	¾"
G	12	1"
H	13.5	1"
I	3.0	½"
J	3.75	¾"
K	4.5	¾"

Pipe Section	Fixture Units	Pipe Size
L	6.0	¾"
M	1.125	½"
N	4.125	¾"
O	5.25	¾"
P	11.25	1"
Q	11.25	1"
R	23.5	1"
S	23.5	1"
T	23.5	1"
U	23.5	1"

*NOTE: When the demand at any point in the water distribution system would exceed the "total demand" you stop adding fixture units and maintain the total demand count to the meter. Section 610.9 states: "No branch piping shall exceed the total demand in fixture units for the system computed from Table 610.3."

Drawing 12

Pipe Section	Appliance(s) Served	Btu/h	CFH Demand	Developed Length	Sizing Length	Sizing Table	Pipe Size
1	Gas Log	80,000	80 CFH	71 ft.	80 ft.	1215.2(1)	3/4"
2	Gas Log/Range	140,000	140 CFH	71 ft.	80 ft.	1215.2(1)	1"
3	Gas Log/ Range/Furnace/Water Heater	340,000	340 CFH	71 ft.	80 ft.	1215.2(1)	1-1/4"
4	Gas Log/Range/Furnace/Water Heater/Clothes Dryer	375,000	375 CFH	71 ft.	80 ft.	1215.2(1)	1-1/4"
5	Gas Log/Range/Furnace/Water Heater/Clothes Dryer/Generator	475,000	475 CFH	71 ft.	80 ft.	1215.2(1)	1-1/2"
6	Range	60,000	60 CFH	69 ft.	80 ft.	1215.2(1)	3/4"
7	Furnace	150,000	150 CFH	69 ft.	80 ft.	1215.2(1)	1"
8	Furnace/Water Heater	200,000	200 CFH	69 ft.	80 ft.	1215.2(1)	1"
9	Water Heater	50,000	50 CFH	67 ft.	80 ft.	1215.2(1)	1/2"
10	Clothes Dryer	35,000	35 CFH	43 ft.	80 ft.	1215.2(1)	1/2"
11	Generator	100,000	100 CFH	30 ft.	80 ft.	1215.2(1)	3/4"

Drawing 13

Pipe Section	Appliance(s) Served by Pipe Section	Btu/h	CFH Demand	Developed Length	Sizing Length	Sizing Table Used	Pipe Size
1	Barbecue	35,000	39 CFH	85 ft.	90 ft.	1215.2(20)	1/2"
1a	Barbecue	35,000	39 CFH	85 ft.	90 ft.	1215.2(1)	1/2"
2	Barbecue/Clothes Dryer	75,000	83 CFH	85 ft.	90 ft.	1215.2(1)	3/4"
3	BBQ/Clothes Dryer/Range	135,000	150 CFH	85 ft.	90 ft.	1215.2(1)	1"
4	BBQ/Clothes Dryer/Range/Log	215,000	239 CFH	85 ft.	90 ft.	1215.2(1)	1-1/4"
5	BBQ/Clothes Dryer/Range/Log/ Water Heater/Furnace	370,000	411 CFH	85 ft.	90 ft.	1215.2(1)	1-1/4"
6	Clothes Dryer	40,000	44 CFH	63 ft.	70 ft.	1215.2(1)	1/2"
7	Range	60,000	67 CFH	61 ft.	70 ft.	1215.2(14)	25 EHD
8	Gas Log	80,000	89 CFH	61 ft.	70 ft.	1215.2(14)	30 EHD
9	Furnace	120,000	133 CFH	49 ft.	50 ft.	1215.2(1)	3/4"
10	Furnace/Water Heater	155,000	172 CFH	49 ft.	50 ft.	1215.2(1)	1"
11	Water Heater	35,000	39 CFH	47 ft.	50 ft.	1215.2(1)	1/2"

Drawing 14

Pipe Section	Appliance(s) Served	Btu/h	CFH Demand	Developed Length	Sizing Length	Sizing Table	Pipe Size
1	Clothes Dryer	35,000	32 CFH	57 ft.	60 ft.	1215.2(8)	1/2"
2	Clothes Dryer/BBQ	75,000	68 CFH	57 ft.	60 ft.	1215.2(8)	5/8"
3	Clothes Dryer/BBQ/Gas Log	155,000	141 CFH	57 ft.	60 ft.	1215.2(8)	1"
4	Clothes Dryer/BBQ/Gas Log/ Furnace/Water Heater/Refrigerator	358,000	325 CFH	57 ft.	60 ft.	1215.2(8)	1-1/4"
5	Barbecue	40,000	36 CFH	56 ft.	60 ft.	1215.2(8)	1/2"
6	Gas Log	80,000	73 CFH	44 ft.	50 ft.	1215.2(8)	5/8"
7	Furnace	150,000	136 CFH	34 ft.	40 ft.	1215.2(8)	1"
8	Water Heater	50,000	45 CFH	55 ft.	60 ft.	1215.2(8)	5/8"
9	Water Heater/Refrigerator	53,000	48 CFH	55 ft.	60 ft.	1215.2(8)	5/8"
10	Refrigerator	3,000	3 CFH	36 ft.	40 ft.	1215.2(8)	1/2"*

* Section 1214.4 - Size of the supply piping outlet for a gas appliance shall not be less than 1/2 of an inch (15mm).

Drawing 15

Pipe Section	Appliance(s) Served by Pipe Section	Btu/h	CFH Demand	Developed Length	Sizing Length	Sizing Table Used	Pipe Size
1	Barbecue	35,000	39 CFH	58 ft.	60 ft.	1215.2(14)	23 EHD
2	BBQ/Clothes Dryer/Range/Gas Log/Furnace/Water Heater	370,000	411 CFH	58 ft.	60 ft.	1215.2(1)	1-1/4"
3	Clothes Dryer	40,000	44 CFH	55 ft.	60 ft.	1215.2(14)	23 EHD
4	Range	60,000	67 CFH	49 ft.	60 ft.	1215.2(14)	23 EHD
5	Gas Log	80,000	89 CFH	56 ft.	60 ft.	1215.2(14)	30 EHD
6	Furnace	120,000	133 CFH	52 ft.	60 ft.	1215.2(14)	31 EHD
7	Furnace/Water Heater	155,000	172 CFH	52 ft.	60 ft.	1215.2(14)	37 EHD
8	Water Heater	35,000	39 CFH	50 ft.	60 ft.	1215.2(14)	23 EHD

Drawing 16

Pipe Section	Appliance(s) Served	Btu/h	CFH Demand	Developed Length	Sizing Length	Sizing Table	Pipe Size
1	Roof Top Unit (RTU) #1	200,000	193 CFH	83 ft.	90 ft.	1215.2(4)	1/2"
2	RTU#1/RTU#2	400,000	386 CFH	83 ft.	90 ft.	1215.2(4)	1/2"
3	RTU#1/RTU#2/Unit Heater/Water Heater/Gas Log	749,900	725 CFH	83 ft.	90 ft.	1215.2(4)	3/4"
4	RTU#1/RTU#2/Unit Heater/Water Heater/Gas Log/RTU#3	949,900	918 CFH	83 ft.	90 ft.	1215.2(4)	3/4"
5	RTU#1/RTU#2/Unit Heater/Water Heater/Gas Log/RTU#3/RTU#4	1,149,900	1,111 CFH	83 ft.	90 ft.	1215.2(4)	1"
6	RTU#2	200,000	193 CFH	74 ft.	80 ft.	1215.2(4)	1/2"
7	Unit Heater/Water Heater/Gas Log	349,900	338 CFH	64 ft.	70 ft.	1215.2(4)	1/2"
8	RTU#3	200,000	193 CFH	57 ft.	60 ft.	1215.2(4)	1/2"
9	RTU#4	200,000	193 CFH	46 ft.	50 ft.	1215.2(4)	1/2"
10	Water Heater	199,900	193 CFH	23 ft.	30 ft.	1215.2(1)	3/4"
11	Water Heater/Unit Heater/Gas Log	349,900	338 CFH	23 ft.	30 ft,	1215.2(1)	1"
12	Gas Log	80,000	77 CFH	20 ft.	20 ft.	1215.2(14)	23 EHD
13	Unit Heater	70,000	68 CFH	17 ft.	20 ft.	1215.2(14)	23 EHD

Drawing 17

Pipe Section	Appliance(s) Served	Btu/h	Developed Length	Sizing Length	Sizing Table	Pipe Size
1	Gas Log/Range/Furnace/Water Heater/Clothes Dryer	375,000	38 ft.	40 ft.	1215.2(30)	3/8"
2	Gas Log	80,000	63 ft.	80 ft.	1215.2(27)	1/2"
3	Gas Log/Range	140,000	63 ft.	80 ft.	1215.2(27)	3/4"
4	Gas Log/Range/Furnace/Water Heater	340,000	63 ft.	80 ft.	1215.2(27)	1"
5	Gas Log/Range/Furnace/Water Heater/Clothes Dryer	375,000	63 ft.	80 ft.	1215.2(27)	1"
6	Range	60,000	61 ft.	80 ft.	1215.2(27)	1/2"
7	Furnace	150,000	61 ft.	80 ft.	1215.2(27)	3/4"
8	Furnace/Water Heater	200,000	61 ft.	80 ft.	1215.2(27)	3/4"
9	Water Heater	50,000	59 ft.	60 ft.	1215.2(27)	1/2"
10	Clothes Dryer	35,000	35 ft.	40 ft.	1215.2(27)	1/2"

PLUMBING FITTINGS EXAMINATION

1 B	14 B	27 D	40 C	53 A
2 C	15 D	28 A	41 D	54 D
3 D	16 B	29 A	42 A	55 A
4 C	17 D	30 A	43 D	56 D
5 A	18 B	31 A	44 D	57 C
6 C	19 A	32 C	45 A	58 B
7 B	20 A	33 C	46 C	59 D
8 D	21 D	34 D	47 A	60 B
9 C	22 C	35 C	48 C	61 D
10 D	23 D	36 A	49 B	62 A
11 B	24 B	37 D	50 D	63 C
12 A	25 D	38 A	51 A	
13 A	26 C	39 D	52 D	